Community-Based Adaptation to Climate Change

As climate change adaptation rises up the international policy agenda, matched by increasing funds and frameworks for action, there are mounting questions over how to ensure the needs of vulnerable people on the ground are met. Community-based adaptation (CBA) is one growing approach that argues for tailored support at the local level to enable vulnerable people to identify and implement appropriate community-based responses to climate change themselves.

Community-Based Adaptation to Climate Change: Scaling it up explores the challenges for meeting the scale of the adaptation challenge through CBA. It asks the fundamental questions: How can we draw replicable lessons to move from place-based projects towards more programmatic adaptation planning? How does CBA fit with larger scale adaptation policy and programmes? How are CBA interventions situated within the institutions that enable or undermine adaptive capacity?

Combining the research and experience of prominent adaptation and development theorists and practitioners, this book presents cutting-edge knowledge that moves the debate on CBA forward towards effective, appropriate and 'scaled-up' adaptive action.

E. Lisa F. Schipper is an Associate at the Stockholm Environment Institute in Stockholm, Sweden.

Jessica Ayers is a Researcher with the Climate Change Group at the International Institute for Environment and Development (IIED), UK.

Hannah Reid is a Consulting Researcher for the International Institute for Environment and Development (IIED), UK.

Saleemul Huq is a Senior Fellow with the Climate Change Group at the International Institute for Environment and Development (IIED), UK.

Atiq Rahman is Executive Director of the Bangladesh Centre for Advanced Studies (BCAS), Bangladesh.

This book provides impressive analysis of community-led processes, based on local priorities, needs, knowledge and capacities, that empowers people to cope with and plan for climate change impacts. It explains why the use of genuine participatory processes is key if adaptation plans are to sit within community priorities. The case studies provide an excellent account of approaches to increasing the resilience of the poor and the most vulnerable communities to the impacts of climate change.

Pa Ousman Jarju, Special Climate Change Envoy of the Gambia, Former Chair of the Least Developed Countries Group to the United Nations Framework Convention on Climate Change, Gambia

As brought out clearly by the Fourth Assessment Report (AR4) of the IPCC and the recently completed Special Report on Managing the Risks of Extreme Events and Disasters to Advance Climate Change Adaptation, the impacts of climate change will require adaptation as a response to the challenge. This book highlights the importance of community-based adaptation measures and provides a wealth of knowledge, such that society across the globe can deal with the impacts of climate change to the extent possible through appropriate adaptation measures rooted in the capacity and strengths of local communities.

R.K. Pachauri, Director-General, Energy and Resources Institute, Chair, Intergovernmental Panel on Climate Change, India

Community-Based Adaptation to Climate Change

Scaling it up

Edited by E. Lisa F. Schipper, Jessica Ayers, Hannah Reid, Saleemul Huq and Atiq Rahman

Routledge
Taylor & Francis Group
LONDON AND NEW YORK

earthscan
from Routledge

First published 2014
by Routledge
2 Park Square, Milton Park, Abingdon, Oxon OX14 4RN

Simultaneously published in the USA and Canada
by Routledge
711 Third Avenue, New York, NY 10017

Routledge is an imprint of the Taylor & Francis Group, an informa business

British Library Cataloguing in Publication Data
A catalogue record for this book is available from the British Library

Library of Congress Cataloging-in-Publication Data
Community-based adaptation to climate change: scaling it up/edited by Jessica Ayers, Lisa F. Schipper, Hannah Reid, Saleemul Huq and Atiq Rahman.
pages cm
Includes bibliographical references and index.
1. Climatic changes–Social aspects. 2. Human beings–Effect of climate on. 3. Community development. 4. Community organization. I. Ayers, Jessica.
QC903.C63 2014
363.738′747–dc23

2013012905

ISBN: 978-0-415-62369-8 (hbk)
ISBN: 978-0-415-62370-4 (pbk)
ISBN: 978-0-203-10506-1 (ebk)

Typeset in Times
by Wearset Ltd, Boldon, Tyne and Wear

Contents

Figures

Tables

Boxes

Contributors

Andrew Adwera holds a MA in Science, Society and Development from the Institute of Development Studies, University of Sussex. He is a research fellow with Research Into Use, working on energy and water security, agriculture and food security as well as capacity building. He has worked at the Africa Centre for Technology Studies in Nairobi and actively participates in national, regional and global processes that aim to shape policies in environment and agriculture.

Diane Archer is a Researcher in the Human Settlements Group at the IIED in London. Her work focuses on urban climate change resilience, with a particular focus on Asian cities. She also has experience of working with community groups in Asian cities on community-driven housing and infrastructure upgrading processes and financial mechanisms.

Jessica Ayers is a Researcher and Policy Advisor working on equitable approaches to managing climate change. At the time of editing, she worked for the IIED on pro-poor climate change planning, evaluation and finance in South Asia and Africa. She holds a PhD in development studies and is a contributing author to the *Fifth Assessment Report of the Intergovernmental Panel on Climate Change*.

Stephan Baas has worked for the Food and Agriculture Organization of the United Nations since 1994. His expertise is climate change adaptation and disaster risk reduction from an agricultural sector perspective. At the centre of his interests are institutional issues related to CCA and DRR at international, national and local levels, including CBDRR and CBA. He has been leading numerous DRR and CCA field projects in Asia, Africa and the Caribbean. In 2011 he coordinated the development of FAO's Framework Program on Adaptation, called FAO Adapt.

Rachel Berger, following earlier careers as an economist, town and country planner, landscape architect and environmental consultant, completed a Masters in Development in 2000. From 2001 until 2012 she worked for Practical Action, a development NGO, on rural livelihoods programmes, before becoming Climate Change Policy Advisor, focusing on support to enable the poorest communities to adapt to climate change. Rachel currently undertakes some writing and consultancy work in the field of climate change adaptation.

Dinanath Bhandari holds an MSc in Forestry and combines specialist skills in different aspects of community-based natural resource management, disaster risk reduction and adaptation to climate change within a broader sustainable development. His work for the last seven years has been to improve resilience of vulnerable communities to adapt to the impacts of climate change and mainstreaming community-based approaches such as livelihood-centred approaches to disaster risk reduction and climate change adaptation to development planning.

Stephane Bonduelle is the Country Director at Concern Universal. He has 18 years of experience in humanitarian assistance and worked with the Red Cross for 15 years, predominantly in management of relief operations. He lived and worked in ten countries before moving to Bangladesh at the beginning of 2007.

Somsook Boonyabancha is the Secretary General of the Asian Coalition for Housing Rights (ACHR), a network of grassroots groups and NGOs supporting urban community-driven development across Asia, based in Bangkok, Thailand. She is a Thai architect and planner who was previously the Director of the Community Organisations Development Institute (CODI), a public organisation which supports community-driven secure housing in Thailand, and also worked for the Thai National Housing Authority.

Maxwell Boykoff is an Assistant Professor in the Cooperative Institute for Research in Environmental Sciences at the University of Colorado, Boulder. He teaches in the Environmental Studies programme and is Adjunct faculty in the Geography Department. Max is also a Senior Visiting Research Associate in the Environmental Change Institute at the University of Oxford. He continues research in areas of climate adaptation, cultural politics and environmental governance, science–policy interactions and political economy and the environment.

Anna Brown is an Associate Director in the Rockefeller Foundation's Asia Office, and manages the Asian Cities Climate Change Resilience Network (ACCCRN), while also covering other urban work. Her experiences include coordinating the MIT-USGS Science Impact Collaborative to improve the use of science in decision-making, leading the Quaker United Nations Office's sustainable development efforts, and research in Asia and Africa. She has a master's degree in city planning from MIT and a bachelor's degree in environmental studies from Brown University.

Pramita Harjati Budihardjo is currently working for Mercy Corps Indonesia as Chief of Party for the SCALE Resilience project, funded by USAID since the end of 2010. The project covers the activity areas of disaster risk management, climate change adaptation, vulnerability and capacity assessments, raising awareness and community resilience pilot projects in four provinces in Indonesia. Previously, from 2009 she was in charge in another climate adaptation project as Project Coordinator of Asian Cities Climate Change Resilience Network project (ACCCRN Indonesia). The project, funded by the Rockefeller Foundation, is focusing on improving the capacity of city stakeholders to identify, develop, pilot, and document climate change resilience activities with potential for replication in other cities in Indonesia.

Maria Rebecca Campos is a Consultant on fisheries economics, and Executive Committee Member of the International Institute of Fisheries Economics and Trade. She started doing consulting work in 1995 as National Technical Expert (Resource Economist) at the Asian Development Bank and United Nations Development Programme. Dr Campos has authored papers and book chapters on risk, adaptation and coping strategies of fishing communities in the face of natural calamities and climate change.

Terry Cannon is a Research Fellow in the Climate Change Team at the Institute of Development Studies, UK. He has a special interest in community-based adaptation (and its limitations), vulnerability analysis and disaster risk reduction. He teaches on climate change and disasters at King's College London, and universities of Copenhagen, Lausanne, Lund, Geneva. He is director of studies at the International Centre for Climate Change and Development in Bangladesh. He is co-author (with Wisner, Blaikie and Davis) of *At Risk: Natural Hazards, People's Vulnerability and Disasters*.

Norberto Carcellar is a priest of the Congregation of the Mission (CM) of St Vincent De Paul, and has varied expertise in community development, community housing and upgrading finance, urban development and disaster interventions, which stems from extensive work organising and empowering communities. He works closely with the Homeless People's Federation Philippines, Inc. (HPFPI), the Philippine Action for Community-led Shelter Initiatives, Inc. (PACSII) and the Asian Coalition for Housing Rights (ACHR).

Gina Castillo is Agriculture Program Strategist at Oxfam America. Prior to this position, she was the global advisor on climate change providing strategic direction for Oxfam's climate change initiatives. Before joining Oxfam America, she was Senior Livelihoods Advisor at Oxfam Novib in the Netherlands. She earned a PhD from York University in Toronto, Canada. She has over 15 years of experience in programme management and research in international development.

Muyeye Chambwera is a Technical Advisor with UNDP Botswana, focusing on sustainable development. He has previously worked for the IIED in London, leading the theme on the economics of climate change, developing and applying economic methods for climate change adaptation that are suitable to developing countries. He has experience undertaking research and policy work in Africa and Asia, focusing on sustainable development, natural resources management and climate change.

Achala Chandani is a Senior Researcher of the Climate Change Group at the International Institute for Environment and Development (IIED), London. Achala has more than ten years of research experience in the field of international law, global climate policy and governance. Her interests are in equity and fairness issues in international climate change negotiations, adaptation to climate change, finance for climate change adaptation and issues related to loss and damage. Achala is a lead author of the chapter 'Climate Resilient Pathways: Adaptation, Mitigation and Sustainable Development' in the *Fifth Assessment Report of the Intergovernmental Panel on Climate Change.*

Sarwat Chowdhury, PhD, works at UNDP Bangladesh as a Climate Change Specialist. Previously she worked as an Energy Consultant and as a Programme Specialist at RBAP, UNDP, New York and was an environment and energy focal point for the Regional Bureau for Asia Pacific, UNDP. Sarwat was also task manager of a Climate Change Adaptation project for the World Bank in Bangladesh and the World Bank Bangladesh's disaster management focal point. She has lived and worked in Bangladesh, the UK, USA and Zambia.

Karina Copen is a Program Officer at Oxfam America, based in El Salvador.

Srijita Dasgupta is a Senior at Mahidol University International College, Thailand, majoring in Environmental Science with emphasis on climate change adaptation and mitigation mechanisms. Her work focuses on the betterment of the marginalised communities near the coastal regions of Bangladesh and elsewhere through better management of ecosystem services and good environmental governance in order to enhance their voice, power and access to resources, with the ultimate goal to strengthen their livelihoods.

Ashvin Dayal is the Rockefeller Foundation's Managing Director for Asia, where he leads several major initiatives, including the Asian Cities Climate Change Resilience Network and SPEED, an effort aimed at promoting renewable energy in rural communities. Before joining the Foundation, he spent 15 years with Oxfam Great Britain in several parts of Asia and the Middle East. Mr. Dayal holds a bachelor's degree from the School of Oriental and African Studies and a Master's degree from the School of Advanced International Studies at Johns Hopkins University.

David Dodman is a Senior Researcher at the IIED, where he is team leader for institutional objectives on 'Cities and Climate Change' and 'Research Quality'. He is the author of more than 30 journal articles and book chapters, and is a College Teaching Fellow at University College London and a lead author on the *Fifth Assessment Report of the Intergovernmental Panel on Climate Change.*

Rada Dukpa works with the Ministry of Health, Royal Kingdom of Bhutan.

Kristie L. Ebi is Executive Director of the Technical Support Unit for Working Group II (Impacts, Adaptation, and Vulnerability) of the Intergovernmental Panel on Climate Change (IPCC). Prior to this position, she was an independent consultant conducting research on the impacts of and adaptation to climate change for more than a dozen years, including on extreme events, thermal stress, food-borne safety and security and vector-borne diseases. She has worked with the World Health Organization, the United Nations Development Programme, USAID and others on implementing adaptation measures in low-income countries.

Jonathan Ensor works at the Centre for Applied Human Rights, University of York, UK. His research focus is on community-based adaptation (CBA), with a particular interest in exploring the relationships between development, adaptation and human rights and the meaning of resilience and adaptive capacity in theory and practice. Prior to joining York, he worked at the international NGO Practical Action, where he was concerned with similar issues and published two books on CBA and development practice.

Richard Ewbank has 20 years of experience (including eight years in Malawi, Zimbabwe and Kenya) focusing on rural development, emergency/conflict response and impact assessment across ten African countries. This was followed by five years with Christian Aid as Climate Advisor, covering primarily Africa, Central America, Central and South Asia. This has involved providing support (training, toolkit development, etc.) to country programmes and partners on resilient livelihoods, with a focus on climate change, sustainable agriculture and renewable energy.

Tim Forsyth teaches at the Department of International Development, London School of Economics and Political Science. He has conducted research on the politics of environmental risk and livelihoods in Southeast Asia, and especially Thailand, since the 1990s. He is interested in community-based adaptation as a growing arena where local development concerns can be integrated with diverse forms of risk assessment.

Jessica Frank is the Climate Change Project Manager at Twin, a UK-based NGO working with smallholder farming cooperatives in developing countries to redress unequal relations through fairer trade. Twin works with farmer cooperatives to support community-based adaptation, with a specific focus on developing best practice in climate-smart production systems. Jessica holds an MSc in Environmental Science from Imperial College London and a BSc in Physics from the University of Bristol.

Fumiko Fukuoka is Senior Technical Advisor and Team Leader for Community Resilience and Sustainability at UNDP, responsible for managing UNDP-GEF's community finance portfolio, including CBA and production landscape management. Fumiko has over 20 years of experience in policy advisory and programming in inclusive sustainable development and partnership development, both at the global and country levels. Previously, she was the Founding Director of the Conservation International Japan Program and Deputy Resident Representative of UNDP in Syria.

Pascal Girot currently holds a position as Regional Climate Change Advisor for Latin America and Caribbean in CARE International, as part of the Poverty, Environment and Climate Change Network (PECCN). Over the past 25 years, Pascal has worked as an international consultant for UNDP, UNEP, IADB, FAO and IUCN. In particular, his work in Central America and the Caribbean has focused on environmental policy, land use planning and protected areas, disaster risk reduction and adaptation to climate change.

Joy Shamake-Guillemot, DrPH, MSc, is an environmental health scientist specializing in climate-related public health policy and programming. As Technical Officer at the World Health Organization she provided technical guidance on climate risk assessment and health adaptation. Joy has worked with a range of health NGOs in Africa, Asia, and Latin America; and consults for WMO, UNICEF, USAID, and WHO on climate adaptation and health.

James Hardcastle is a nature conservationist. He has spent 20 years working internationally, including the Indian Ocean, the Pacific and Southeast Asia. His efforts have helped develop practical, hands-on approaches to community development that are nature-based, cost-effective and deliver results. He works to represent the lessons learned from conservation efforts that can inform national policies and global development objectives, such as in a role as an advisor to island nations at the UN Climate Change Convention negotiations. He currently works for the IUCN's Global Protected Area Programme, based in Switzerland.

Zarina O. Hipolito has a degree in journalism and community development, and an extensive background in development communication and electricity cooperative training. She works as a Documenter for PACSII and the HPFPI in the Philippines, and prior to that worked as a Research Associate with the Community-based Coastal Resources Management Resource Centre.

Saleemul Huq is Director of the International Centre for Climate Change and Development at the Independent University, Bangladesh, and a Senior Fellow at the IIED in London.

Albon Ishoda is a Marshallese naturalist and community development champion. He is the Chief Executive of the Marshall Islands Conservation Society, based in Majuro. Albon has contributed greatly to raise awareness of the challenges and opportunities that island and atoll communities face under a future of climate change in his country, and across the Micronesia and Pacific region. He has led efforts to deliver climate adaptation solutions that can work for remote atoll communities. Albon has represented the Marshall Islands at global events and at UN climate negotiations.

Aminul Islam, PhD, has been associated with UNDP Bangladesh since 1996 and served different positions in the Country Office, starting as the Sustainable Development Advisor and serving as the Assistant Country Director to lead the Climate Change, Environment and Disaster Management Cluster. He currently holds the position of Senior Advisor. Dr Islam also served as the Senior Technical Advisor and Team Leader of the Water Disaster Risk Reduction Project in Central Vietnam, supported by UNDP Vietnam. He also taught environment and climate and research methodology both at under- and postgraduate level in different public and private universities.

Helen Jeans is the Climate Change Adaptation Network Coordinator at WWF-UK. She has a background in law, international development and human ecology and has spent over ten years working in the Pacific Islands, including on climate change adaptation. She has a keen interest in learning-based approaches to climate change adaptation.

Lindsey Jones is a Senior Research Officer at the Overseas Development Institute. His background is in international development and environmental policy. His research interests are in the linkages between climate change, adaptation and development. Lindsey has previously worked with the United Nations Development Programme (UNDP) and World Food Programme (WFP). He has an MSc in Environmental Policy from the University of Oxford and a BSc in Environmental Geography and International Development from the University of East Anglia.

Iqbal Kabir works in the Climate Change and Health Promotion Unit, Ministry of Health and Family Welfare, Bangladesh.

Bettina Koelle is the Director of Indigo Development & Change. She has been working in development for the past 15 years, focusing on trans-disciplinary research and participatory methods. She is involved in climate change adaptation working with rural communities, training, social research, participatory GIS and organisational development. Bettina is currently also undertaking her PhD studies in the field of climate change adaptation – focusing on how to increase problem-solving in communities in the adaptation process to increase resilience to climate change and climate variability.

Simon Levine spent many years working for NGOs in Mozambique, Cambodia, Tanzania and Burundi, before working as a consultant based in Uganda for nine years. He specialises in livelihoods and vulnerability analysis, land rights and early response in humanitarian crises.

Jo Lofthouse is based in London and was the coordinator of the Africa Climate Change Resilience Alliance. Her interest in community-based adaptation focuses around the role that civil society organisations, local and national government can play in supporting communities to adapt to climate change. She has extensive experience in sub-Saharan Africa, working for DFID and Oxfam GB.

Denise Loga, MSc, is a Climate Change and Sustainable Development Consultant, currently consulting for the Ecological Economic Research Institute. She has previously worked with the IIED, providing technical support to projects on economics of adaptation in the water sector. As a Research Associate for the United Nations Environment Programme in Beijing, she supported International Ecosystem Management Partnership projects related to climate change adaptation, eco-investment and green economy.

Eva Ludi is a Research Fellow at ODI and a Research Associate at the Centre for Development and Environment of Bern University, Switzerland. She has over 15 years of experience in international policy-oriented research in Asia and Africa. Her work focuses on natural resource management, rural livelihoods and environmental conflicts. Recently, she has been focusing on water and food security links and on adaptive capacity in the context of strengthening resilience of vulnerable people. Eva has also taught at university level for several years.

Mazan Malkawi, MSc, is a Technical Officer at World Health Organization Centre for Environmental Health Activities in Jordan. He is an Environmental Engineer with extensive experience in managing water and sanitation related health risks.

Jessica Mercer has over nine years of experience in the International Development sector across Asia-Pacific, Africa and Latin America, where she has specialised in disaster risk reduction (DRR) and climate change adaptation (CCA). Having worked with NGOs, UN agencies and academia, Jessica has now set up Secure Futures consultancy (www.secure-futures.net) with a small group of social and physical scientists focused on risk reduction for communities worldwide.

Paul Mitchell is Save the Children's Senior Climate Change Advisor, responsible for integrating climate risk and adaptation into development and humanitarian programmes across the Asia-Pacific. Previously, Paul managed the Australian aid programme's International Climate Change Adaptation Initiative. He has developed, implemented, monitored and evaluated adaptation and development projects and strategies at local, national and regional scales in the Pacific, South and Southeast Asia, Africa and the Caribbean. He lives in Hanoi, Vietnam.

Thuan Nguyen, MA, is Health and Care Program Manager, International Federation of Red Cross and Red Crescent Societies Vietnam Delegation, where she has experience implementing community health programs in Vietnam.

Charles Nyandiga, PhD, is working as an advisor in community-based climate change adaptation, managing two CBA projects covering 48 countries. His expertise includes climate change and natural resources management. His experience includes, working as a forest researcher and practitioner with government, bilateral organisations and the United Nations Development Programme. Currently, he is the Technical Specialist for the GEF SGP Sustainable Forest Management, Land Degradation and Climate Change Adaptation at UNDP, based in New York.

Martin Obermaier is a climate change adaptation expert, and a moderator of the Evidence and Lessons from Latin America (ELLA) network. ELLA is a knowledge-sharing and learning platform that shares knowledge of recent Latin American experiences on selected economic, environmental and governance issues, supports learning between Latin American, African and South Asian countries and provides a networking platform for organisations and individuals to link to Latin America.

Judy Oglethorpe leads the Hariyo Ban Program, which aims to reduce adverse impacts of climate change and threats to biodiversity in Nepal. Previously she managed WWF-US's Climate Adaptation Program, and before that, conservation programmes linked with health, population, poverty, livelihoods and conflict. She worked for 14 years in Southern/East Africa in biodiversity, forest and wildlife conservation and CBNRM. Her qualifications are MSc Environmental Management, University of London; and BSc Ecological Science, University of Edinburgh.

Agnes Otzelberger currently works to develop and strengthen CARE International's community-based adaptation work in Africa, and to enhance integration of social and gender equity into CARE's global climate change work. Agnes is a political scientist by training and holds an interdisciplinary MA from the Institute of Development Studies at Sussex University. She has worked in research, policy processes and programming on climate change, disasters and food security across Africa, South East Asia and South Asia.

Marcus Oxley has over 20 years of disaster risk management experience gained within the civil society sector, where he has held responsibilities for supporting, facilitating and implementing a range of disaster risk management activities across Africa, Asia and Europe. In June 2007 he became the Founding Chair and subsequent Director of the Global Network of Civil Society Organisations for Disaster Reduction – a major international alliance of large and small civil society organisations committed to working together to put the voices and concerns of vulnerable people at the heart of global disaster risk reduction policy and practice. Marcus is a civil engineer by training and has a master's degree in Project Management.

Joanna Phillips is Head of Adaptation at WWF-UK, and before that was Head of International Development Policy at the Royal Society for the Protection of Birds (RSPB). She works at the environment and international development interface, drawing on over 12 years of experience in sustainable development policy, advocacy and stakeholder engagement at international, national and local levels, most recently to address climate change, building resilience and adaptive capacity in people and ecosystems.

Indrani Phukan was Senior Programme Officer, South Asia regional office of Christian Aid. She was managing the Strengthening Climate Resilience Programme being implemented in India, Bangladesh, Sri Lanka and Nepal until October 2011. She is now with Intercooperation India as Programme

Coordinator for Climate Change, with the responsibility to develop and evolve a climate change portfolio.

Atiq Rahman, the Executive Director of Bangladesh Centre for Advanced Studies, is a well-known scientist of climate change and sustainable development. He was a lead author of the *IPCC Assessment Reports*, which received the Nobel Peace Prize in 2007. He was also awarded the highest UN award on environment, 'The Champion of the Earth 2008' by the UNEP for the Asia-Pacific Region. He leads BCAS and IIED teams for organising the community-based adaptation conferences.

Sudip K. Rakshit is a Canada Research Chair (Tier 1) in Bioenergy and Biorefining Processes at Lakehead University, Thunder Bay, Canada. Earlier he was working at the Asian Institute of Technology (AIT), where he served as the Vice President of Research and initiated a Centre of Excellence in Sustainable Development in the context of climate change. He has considerable experience in South and South East Asia in community-based adaptation for agriculture and water management.

Jason Rayos Co is an alternative lawyer working in the Philippines with marginalised sectors to achieve their development goals through creative use of the law. His interests lie in legal and allied research, documentation of community-driven processes, upgrading and socialised housing, organisational development, popular education and policy research, and reform advocacy work. He is currently working in the public sector, and is an active member of the Board of PACSII and a volunteer for the HPFPI.

Hannah Reid is a Consulting Researcher, currently working with the Climate Change Group at the IIED in London. She has over ten years of experience working on climate change, with particular focus on how best to help those who are most vulnerable cope with its impacts, and the use of ecosystem-based approaches to help people adapt.

Anna Ricoy has been working for the Food and Agriculture Organization of the United Nations in Rome for the past five years, before recently joining One-World in Cape Town. Her expertise lies in climate change adaptation and disaster risk reduction in the agriculture sectors, including the strengthening of institutional and technical capacities for community-based adaptation. Her work has focused on Latin America and the Caribbean and in Western Africa.

Tine Rossing is a freelance consultant. Prior to this she worked for CARE, the UNDP, World Bank, DANIDA and WRI. Tine has focused extensively on developing integrated approaches to community-based adaptation and participatory M&E that combines ecosystem concerns, DRR, livelihoods and governance aspects. Tine holds an MA in International Relations/Economics from John's Hopkins SAIS and has 19 years of experience in climate change adaptation, biodiversity conservation and poverty alleviation from working in 27 countries on four continents.

E. Lisa F. Schipper is an Associate at the Stockholm Environment Institute in Stockholm, Sweden. Her research specialty is adaptation to climate change, socio-cultural vulnerability to climate change and natural hazards and links between sustainable development, adaptation to climate change and disaster risk reduction in Latin America, Africa and Asia.

Katarina Soltesova is a doctoral student at the Centre for Urban Sustainability and Resilience, based at University College London. The focus of her work is urban community-based disaster risk reduction and urban development. She is presently working in West Africa.

Janot Mendler de Suarez is Visiting Research Fellow, Boston University Pardee Center for the Study of the Longer-Range Future. He was instrumental to the development of GEF-IW:LEARN, the Global Environment Facility's International Waters knowledge-sharing programme involving 182 countries in participatory learning and dialogue. Pioneering new methods for community-based adaptation, recent game design partners include the UN World Food Programme, Oxfam, Rockefeller Foundation, Red Cross/Red Crescent Climate Centre, Africa Climate Change Resilience Alliance and the World Bank.

Pablo Suarez is Associate Director for Research and Innovation at the Red Cross/Red Crescent Climate Centre, and visiting fellow at Boston University. He has consulted for WFP, UNDP, Oxfam and about 20 other international humanitarian and development organisations, working in over 60 countries. He develops participatory tools for climate risk management, linking global knowledge to action for vulnerable communities. Pablo holds a PhD in geography, is based in Boston and covers Africa and the Americas.

Sarah Wiggins has 20 years of experience in the UK and international voluntary sector, from community settings to international policy arenas. Her work has focused on participatory methodologies and justice for people living in poverty. For six years this focus has been applied to climate change adaptation practice and policy and her work has included writing the Tearfund CEDRA tool to help local NGOs in developing countries to manage risks from climate change and environmental degradation.

Kirsty Wilson was Ethiopia Project Manager of the Africa Climate Change Resilience Alliance with Oxfam until December 2011. During 2012 she worked with Oxfam on humanitarian, agricultural, pastoral and climate policy and advocacy. She is currently studying for her MSc in Environmental Change and Management at Oxford University.

Ben Wisner works with ABUCH, University College London (www.abuhc.org) and GNDR (www.globalnetwork-dr.org). He studies complexity, local knowledge and interactions among livelihoods, governance, natural environment and risk. He taught in Africa, the USA, UK and Switzerland for over three decades, while researching in Africa, Asia and Latin America. He now consults, advises and writes. Ben is lead author of *At Risk: Natural Hazards,*

People's Vulnerability and Disasters (2004) and lead editor of *The Routledge Handbook of Hazards and Disaster Risk Reduction* (2012).

Leena Wokeck is the Director of the CSR Asia Center at AIT in Thailand and an Executive Director of CSR Asia. Her expertise is in responsible and inclusive business strategy, transparency and accountability, and private sector engagement on various aspects of sustainable development, including climate change adaptation. She leads research and engagement projects and works on developing and delivering training and executive education programmes, including the joint CSR Asia-AIT Professional Masters Program in CSR.

Helena Wright is a Doctoral Researcher in Environmental Policy and Technology at the Centre for Environmental Policy, focusing on climate change finance. She has contributed to publications, articles and books on climate change mitigation, adaptation and technology transfer, and has worked in collaboration with the UK Government Department for International Development, Friends of the Earth International and the International Institute for Environment and Development.

Acknowledgements

The editors would like to thank all the authors and co-authors for their hard work and also acknowledge with thanks the advice and support received from Ian Burton, R.K. Pachauri and Pa Ousman. We would also like to thank Mozaharul Alam and S.V.R.K. Prabhakar for insightful discussions. Particular thanks are due to the co-sponsors, NGOs, researchers, government officials, students and other participants who have attended and supported the annual Community-Based Adaptation conferences over the years. The International Institute for Environment and Development and the Bangladesh Centre for Advanced Studies have been working to promote the issue of community-based adaptation through these annual conferences (and other research and advocacy activities) for several years, but ultimately the success of any event depends on the people who attend. Thanks are also due to Earthscan and Routledge staff whose patience and encouragement have ensured the publication of this book, namely Helen Bell, Helena Hurd and Louisa Earls.

Part I

Key concepts for community-based adaptation

1 Upscaling community-based adaptation

An introduction to the edited volume

Hannah Reid and E. Lisa F. Schipper

This book is based on discussions and material presented at the fifth International Conference on Community-Based Adaptation (CBA) held in Dhaka in February 2011. Rather than publishing a set of commissioned papers presented at the conference, however, the book consists of a set of chapters produced by 'writing teams' who have focused on the subject areas identified as being central to CBA at the conference. Much of the material in this book is based on discussions held at the conference and also descriptions of experience with CBA provided by the many practitioners working in the field on CBA. By using writing teams, we hope to capture some of this more experiential knowledge on CBA, which is substantial but not widely written about in traditional scientific literature. This approach will, we hope, help the book capture a wider ranging set of emerging views and knowledge about CBA, and inform important processes such as the Intergovernmental Panel on Climate Change, which has a strong effect on policy, about both the latest science and best CBA practice on the ground. This balance of both scientific and practical knowledge on CBA is essential if any efforts to upscale CBA are to meet with success.

CBA has emerged as one of the most rapidly expanding bodies of knowledge in the greater field of adaptation to climate change. From the outset, the appeal may seem intuitive: CBA embodies adaptation practice that is small-scale, place-based and often grassroots-driven, engages development practitioners and development approaches, and most importantly is community-based. Thus, it takes into account recommendations that have emerged from nearly two decades of research and practice on adaptation. The inclusiveness and lesson-based approach of CBA has clearly hit home with many people; the growing body of researchers, practitioners and community members interested in this approach to adaptation is a testament to the fact that many of us now acknowledge its potential in the fight against climate change. The number of publications and websites on CBA is increasing, and the annual conference on CBA run by the International Institute for Environment and Development and its partners attracts people in ever increasing numbers. A growing body of people clearly feel the concept has merit as a discrete arena for focusing efforts and channelling resources.

Since CBA appears so successful, there is a desire to replicate and upscale this community-friendly approach. But can this be done? Can the main characteristics of CBA actually be upscaled and their advantages maintained? This volume seeks to explore precisely that question through a number of themes. Naturally, some of the key questions explored in the collected chapters cover issues such as how successful CBA really is, and what makes CBA unique. Reflection on this topic also begs the question: is CBA really so 'new'? Is it not just the linking together of two disciplines and research communities – climate change science and local development – whose paths have previously not overlapped to any great degree? Are there dangers in making CBA yet another intellectual silo to meet our own self-serving development agendas and further obscure realities on the ground with, as Professor Bob Pokrant calls it, 'the abstractions of climatocracy'? Or are we just reinventing our ideas with a view to attracting additional multilateral, bilateral donor and national climate change funding? While development agencies may differentiate between disaster risk reduction, climate change adaptation and poverty alleviation, doesn't all of this converge at the household level? A woman in her own house is, after all, her very own minister for planning, health, education, environment and fiscal management. Reid *et al.* (2009) argue 'There is increasing recognition that, for many communities facing frequent hazards, poverty, disasters, and climate change adaptation are closely linked and cannot be viewed in isolation from one another.' At a certain level, that is precisely what CBA does so well – it takes a holistic picture of a development–climate change interaction and offers a holistic solution, simultaneously opening itself up for critiques about how this actually differs from other types of sustainable, local development.

Despite this, a growing body of people see merit in looking at CBA as a discrete issue. Carving out CBA from other discourses provides a way to help us understand what we mean by the term. It also provides us with an opportunity to develop common platforms so that those interested in learning more have places they can go to do this, like-minded colleagues they can share information with while there, and a common set of terms and principles with which to ensure people are talking about the same thing and hence better understand and deepen existing knowledge. We know that adaptation is a long-term process, and that most field sites have no baseline data, counterfactuals or control sites, so in order to assess whether adaptation initiatives have been successful we will need to monitor activities over several years. Without this learning, we will not be able to claim the success or otherwise of such initiatives, and the basis for upscaling them will therefore be limited. Greater learning and understanding will lead to greater effectiveness.

But there is a danger here: without a strong understanding of what CBA is, what it is not, and how best to do it, the concept could shortly join many others in the oversubscribed graveyard of development approaches that have been tried and found wanting (or not tried properly but still found wanting). In the rush to promote this seemingly wholesome approach, we need to be able to clearly explain and defend the logic behind it. This will serve both to provide information and tools to those working to help the poor cope with climate change impacts and scale-up successful approaches, but it will also provide defensible alternatives to other approaches to tackling climate change – such as the construction of large sea-walls or dams – that provide measurable outcomes, impressive political kudos and great photo opportunities, but largely neglect the needs and priorities of those most in need.

There are important lessons to be learned for CBA from other development discourses, particularly the disaster risk reduction community, the literature on 'participation' and also the community-based natural resource management (CBNRM) arena (Dumaru 2010). In its heyday, CBNRM was heralded as an alternative approach to the kinds of conservation activities favoured at the time involving national parks, armed rangers, fences and the separation of natural resources deemed to be of conservation value from the local people who had hitherto relied on them (IIED 1994). Both CBNRM and CBA are 'bottom-up' as opposed to 'top-down' in their approach. In other words, they take the community as the entry point for action and inquiry, rather than using a model-based problem statement about the situation as their starting premise. Yet, amidst the debate about the merits or otherwise of CBNRM, Marshall Murphree made the following important observation. He stated that community conservation 'has to date not been tried and found wanting; it has been found difficult and rarely tried' (Murphree 2000: 12). Other scientists writing at the time also observed that, in practice, approaches labelled as 'community-friendly' varied in their level of anthropocentricity and the degree to which genuine participation occurred (Adams and Hulme 2001).

Ensuring that lessons from CBNRM, its relative community-based disaster risk management (CBDRM) and other participatory development practices are well incorporated into CBA is necessary before we can think about scaling up CBA, in order to ensure mistakes are not repeated. One pertinent lesson from CBNRM that those interested in CBA can learn is that it is important to clearly define the concept of CBA. At the six international conferences on community-based adaptation to climate change held to date, it has been apparent that many organisations are seeking to opportunistically label their work as CBA in order to capitalise on some of the advantages that may result. Such re-packaging of 'old wine in new bottles' in this way is unwise. Just as many approaches labelled as CBNRM were discredited and criticised, when in reality they were top-down and externally initiated and not genuine CBNRM at all, the concept of CBA may be discredited by approaches labelled as CBA, which then 'fail' but which were actually not CBA in the first place. When multiple definitions and concepts are used to mean the same thing, it also suggests the lack of a real community of knowledge.

As with CBNRM, the practice of CBA also needs more analytical rigour in order to assess whether it really works or not. Despite the urgency in terms of action that climate change scenarios tell us is needed to tackle this global threat, CBA must not try to run before it can walk. It will take time to gather the robust evidence needed to objectively explain the merits and also the limitations of CBA. CBA as a concept may have energy and momentum associated with it now, but in view of the changing winds of donor funding and trends in development assistance – and climate change is a very fast-moving and political arena at present – a stronger evidence base is needed if it is to stand the test of time. Many of those currently involved in CBA activities are practitioners working in or close to the field, but the skills of these people need to be complemented by a degree of analytical rigour and objective analysis about the genuine merits and challenges associated with CBA. Without this analysis the merits and challenges of CBA will be obscured, and the concept could lose credibility before it has had the chance to genuinely prove itself or otherwise.

Only when we have clear answers to the questions of 'what is CBA?' and 'what evidence is there demonstrating whether it works or not' can we think about how best to scale-up CBA. The following sections therefore address these issues before addressing the issue of scaling up.

What is CBA? What isn't CBA?

The 'A' in Community-Based Adaptation

While we are still learning what CBA is exactly, we are now getting better at working out what CBA is not. A development project helping communities cope with floods or saline water intrusion in their cropland, for example, is not CBA unless it can be demonstrated that the floodwaters or saline water intrusion causing the problems are a result of climate change as opposed to regular climate variability.

The emphasis on the 'A' of 'CBA' and what this means is important here. Strong scientific or local knowledge about the past, current and expected climate change impacts is needed in order to make the assessment of whether or not an activity can be described as adaptation or not. This is a challenge on the ground, where in most cases it remains difficult to categorically attribute any one climatic event such as a flood or storm directly to climate change. In most cases scientists can only say that we are more or less likely to see such events in an area due to the impacts of climate change. Likewise, climate models can still only identify which parts of the world are likely to be more exposed to climate changes, and at a geographic resolution or timescale that has little relevance to the day-to-day planning conducted by poor farmers in Mali needing information on what crops to plant and when.

Both scientific and local knowledge can contribute to the understanding of whether the risks of climate change impacts are likely to increase or not, and this understanding of risk and uncertainty needs to be built into any genuine CBA activity. Incorporation of this knowledge is one of the key ways in which CBA is different from development as we know it. Retrospectively labelling a development project as CBA when no observed climatic events can be objectively attributed to climate change and the project was not planned based on existing assessments of where climate change risks might significantly increase is repackaging, and therefore unwise, even if well-intentioned.

Pinning down what we mean by the 'A' of CBA also involves addressing the issues of whether we mean adaptation to first-order climate change impacts themselves, such as storms, floods or sea-level rise, or whether this also includes adaptation to various local-, national- and global-scale responses to climate change. For example, would local communities adapting to the effects of newly introduced carbon offsetting policies that incentivise tree-planting on communal land count as CBA? Or shifts in economic policies such as the subsidisation of biofuels or promotion of low-carbon development models as a national response to climate change that affect food and energy prices for local communities? Or even local responses in one place to adaptation responses (such as nearby dam construction) nearby? The authors feel that these second-order adaptation responses are outside what should be classed as CBA, but the question is still valid.

Process, not just outcomes

Reid *et al.* (2009: 13) explain that 'Community-based adaptation to climate change is a community-led process, based on communities' priorities, needs, knowledge, and capacities, which should empower people to plan for and cope with the impacts of climate change.' This definition places an important emphasis on the *process* by which activities are undertaken in order to determine whether or not they can be called CBA. There are lessons to be learned here, again, from CBNRM, where several authors observed that activities labelled as CBNRM were often in fact top-down in nature, perhaps only involving the community in a consultative fashion with a view to meeting pre-conceived objectives.

CBA should not just be work implemented *at* the level of the community, such as building a well in a local village to combat water shortages; rather, work should be *centred around* the priorities and processes chosen by the community. Good CBA should be self-mobilised, whereby people participate by taking initiatives independent of external institutions to change systems (Pretty *et al.* 1995; Adnan *et al.* 1992; Reid *et al.* 2009; Dodman and Mitlin 2011). Good CBA should not prioritise the interests of outsiders over those of the community, and experience from fields such as natural resource management and disaster risk reduction shows that if CBA is to be effective and sustainable, it must 'draw on the knowledge and priorities of local people, build on their capacities, and empower them to make changes themselves' (Reid *et al.* 2009: 23).

Such people-centred approaches do not often sit comfortably with the dominant professional behaviour and institutional structures of industrialised countries. Outsiders must become facilitators and co-learners, not 'teachers' or 'experts'. We saw with CBNRM how wildlife guards struggled with participatory approaches, and we are seeing this again now with climate change experts who are recognising the need to work more 'on the ground' but who, with no solid grounding in the social sciences, struggle to do so effectively. Participatory processes need time to develop and they need flexible funding. They do not fit with the pre-determined calendars, budgets and outputs demanded by governments and most donor organisations.

It is also important to be clear about exactly who we mean when we talk about 'the community' in the context of CBA. Few would argue that 'the community' in the context of any CBA work is anyone but the group of individuals who are affected by climate change impacts and who are working with (or without) external intervention to cope better with these impacts. In practice, however, the priorities, needs, vulnerability and capacities of people within any one community vary tremendously, and it is not always clear who will or should benefit from any CBA initiative. The impact of a flood may be completely different for the men and women in a community: the women's primary concern may be the safety of the children who cannot swim and the issue of how they will get clean drinking water for their family. The men, by contrast, might prioritise how the floodwater will affect their crops or commute to work. So should the needs of the men or the women be prioritised? In any community there will be the 'haves' and the 'have-nots' who vary tremendously in their capacity to undertake adaptation activities, and those with voices (perhaps local chiefs or powerful elites), and those without them (often women and children) who struggle to make their priorities and concerns heard. Should CBA target the 'have-nots' and the voiceless, and if so, how? Or should CBA necessarily be inclusive of everyone in a community?

Currently, several CBA initiatives are quite simplistic in the manner in which they approach the notion of 'community'. This may be because the initiatives are implemented by meteorologists, climate change scientists, or conservation biologists, whose strengths in the social sciences are outweighed by those in the physical sciences. This is an inevitable result of working in such a new and interdisciplinary field, and one which CBNRM has long experience with. But it requires those working in this field to have an open mind and a willingness to learn about ways of working and thinking that may be traditionally outside their comfort zone. Social scientists must learn about the 'hard' science of climate change, the notion of risk and uncertainty and the difference between climate variability and change. Social scientists must learn about participatory processes and how best to prioritise local needs.

Much has been written about the problems of defining who a 'community' is (see, for example, Kepe (1999) for a particularly clear case study), and this introduction will not dwell any longer on the issue, but it is important to acknowledge that identifying who exactly we mean by 'the community' is not as easy as it seems and that we need to be clear about this when working on CBA. It is known that climate change impacts will affect the poorest and most vulnerable the most (Huq *et al.* 2006), so it seems logical that it is these people who should be the focus of any CBA initiative. But ensuring that a CBA intervention benefits the poorest and most vulnerable rather than reinforcing existing power differentials and supporting the priorities of entrenched elites is a challenge that development practitioners have been grappling with for many years, and one for which there are no easy solutions (see, for example, Malla *et al.* (2003) for a good example of this in community forestry groups in Nepal). Furthermore, CBA activities are not necessarily independent of the wider policy context in which they occur, meaning that if those policy contexts change or fail, community-based initiatives may also topple (Schipper 2009).

Other publications provide better direction to those wishing to learn more about the process of how to do good CBA (e.g. Reid *et al.* 2009; Magee 2012), and the range of participatory approaches that can be adopted and pitfalls to avoid. These include, for example, the importance of using local languages and of explaining climate change using both scientific and local forms of knowledge. This volume instead focuses more on the theory of CBA and the context in which it is applied in national and international policy-making and planning for climate change.

Scaling up CBA

Most CBA initiatives to date are small stand-alone activities that operate at a localised scale. As has been demonstrated by the rising interest in presenting material at the annual international CBA conferences, such activities are proliferating rapidly. Interest in how to learn more broadly from the wide variety of CBA activities being initiated on the ground, and share lessons with a view to reaching a greater number of those affected by climate change, is strong. Working out how to actually scale up CBA is the purpose of this book, and ensuing chapters look at scaling up in the context of health, disaster risk reduction, agriculture, urban areas, gender, risk assessment, economics and financing mechanisms, tools and learning.

Uvin and Miller (1996) distinguish between four types of scaling up in the context of non-governmental organisation (NGO) interventions: quantitative, functional, political and organisational. Quantitative scaling up is the geographical spread to more people in the same sector or area. This commonly occurs when programmes get replicated in nearby areas or when one programme expands to benefit more people in a specific area. Functional scaling up works by expanding the scope of an activity – for example, if an agricultural programme expands its focus to include climate change adaptation. Political scaling up refers to expansion through efforts to broaden the political process in which a programme operates, and organisational scaling up means expanding the institutional base implementing the intervention, either by expanding the institution itself or by involving other institutions. This can be both horizontal (involving similar institutions) or vertical (involving local, regional, national or even supranational institutions). In the context of climate change initiatives, Pelling (2011) notes that there are

> three modes through which local innovations can come to influence the wider regime: (1) replication: the horizontal reproduction through multiple, small initiatives, (2) scaling-up: the expansion of individual initiatives as they attract more participants, and (3) mainstreaming: the absorption of innovations into dominant policy and practice.

These definitions of scaling up are helpful, but in this book 'scaling up' is used to mean something more generic. It simply refers to any activity that could move CBA activities from a collection of local projects to something which would reach a much wider spectrum of vulnerable people. This is captured well by the World Bank (2005) definition of scaling up, developed in connection with the 2004 Shanghai conference on scaling up: 'Scaling up means expanding, adapting and sustaining successful policies, programs or projects in different places and over time to reach a greater number of people.' This makes clear that scaling up is not only about projects, but also programmes and policies.

In summarising discussions at the fifth CBA conference, professor Ian Burton identified five key points to consider in the context of scaling up CBA:

1 'Outscaling' by expanding the CBA movement outwards from its current set of theorists and practitioners.
2 Linking upwards, particularly to local, regional and national levels of government.
3 Integrating CBA into wider development planning.
4 Strategising effectively to focus efforts in light of the limits of adaptation.
5 Mitigation: an essential response to the climate change problem not to be left out of any discussion about how to respond best to climate change.

Using these five points as a starting point, the next sections explore what we mean by scaling up CBA and how best to do it.

Outscaling

Currently, those involved with the practice and theory of CBA are quite a small community of development, environmental and climate change professionals. And of course the local people they are working with. As Reid *et al.* (2009: 27) note,

> The theory and practice of CBA are still in their infancy. Both are likely to grow very rapidly, however, as needs increase as a result of intensifying climate change impacts and as interest in and support for adaptation grows at national and international levels.

Outscaling in this context means expanding the CBA movement outwards at the local level to reach more people. Funding is increasingly available to support adaptation activities, both through the international climate change regime, bilateral and multilateral aid programmes, and also national governments. The government of Bangladesh, for example, put approximately $45 million from the national exchequer into a new multi-donor trust fund for climate change under the 2008/2009 national budget (Ayers *et al.* 2009). But the availability of money will not be the only driver of this outscaling. Many of those showing a keen interest in CBA are driven by a desire to improve the rights and livelihoods of vulnerable people, rather than any monetary benefit. Increasingly, communities themselves are starting to notice that 'something is different' in the weather patterns they are experiencing compared to those in the past.

The emphasis on CBA as being as much a process as a science is important here. Steele *et al.* (2008) acknowledge that 'Scaling up implies adaptation, modification and improvement, not just replication. Thus scaling up applies not only to a particular technology and techniques, but more importantly to principles and processes.' This is particularly important in the context of CBA, where, if we accept that the concept itself is defined in part by the participatory processes by which it is undertaken, it is obvious that a 'blueprint' approach to outscaling will never work. What might be good CBA practice in one area could well not be appropriate for replication in another, despite external appearances. It will be up to the community themselves to determine exactly what adaptation activities are a priority for them in the local context. In some instances this might be nothing at all, if communities do not have the capacity to participate or have prioritised other issues. As Reid *et al.* (2009: 27) note, 'To be successful, community-based adaptation programmes will need to ensure that communities are able to participate in identifying priorities, both local and regional, and in planning, implementing, monitoring, and reviewing adaptation.'

A flexible 'learning by doing' or 'social learning' attitude can help accommodate this focus on process as opposed to outcomes (Huq and Reid 2007; Collins and Ison 2009; Dumaru 2010). But it is also important to factor in the need for measuring how effective any activities undertaken have been. This requires a good assessment of baseline starting conditions, either by external actors or the community themselves, followed by accurate recording of changing environmental conditions, community responses and resulting benefits and costs. It also needs a clear articulation of the aims of any activity in order that progress towards those aims can be ascertained. In the context of outscaling, while a community-led approach based on local needs, knowledge and experience should be central to any strategy pursued, it is also important to account for the knowledge and skills that may also exist outside the community, and the need for a clear set of goals and strategy for meeting and measuring progress towards these goals.

Linking upwards

Most CBA activities to date are initiated by non-government organisations, and often with external funding. But if CBA is to be more than a series of isolated initiatives, it needs to be systematically mainstreamed into government structures, policies and planning processes. Non-government organisations come and go, but government structures are more permanent, even if individuals staffing them move on. Embedding existing and future CBA projects in local, regional and national government structures will support the widescale replication of projects, as they will be systematically supported or even 'rolled out' as part of government programmes. It may also protect the programmes from stakeholders who see it as a threat or do not support its aims. As Reid *et al.* (2009: 28) note,

CBA 'programmes should provide support and link communities to relevant decision-making institutions. They will also need to build the capacity of local organisations and local governments to enable them to effectively take part in decision-making processes.'

It is only governments rather than NGOs (with notable and important exceptions such as the Orangi Pilot Project in Pakistan, or the Bangladesh Rural Advancement Committee – BRAC – in Bangladesh, both of which have taken over many of the roles usually tasked to governments) that generally have the capacity to provide the sustained mainstreamed policy and financial support for larger programmes. Donor support, by contrast, is more volatile. Governments also have a role to play in providing sustained support for capacity development; in other words ensuring that those individuals or institutions involved in upscaling have the skills and resources needed to do this, although this is also a role that donors and other external actors can work in partnership with government bodies and communities to fulfil.

Mainstreaming CBA into government policies and programmes will likely be challenging if efforts to upscale CBNRM in China provide a likely measure of success. In discussing the main problems that needed to be overcome to scale-up CBNRM in Guizhou Province, the researchers concluded that 'the challenge remains: how to change institutional arrangements, policy-making mechanisms and daily practices of government in order to create space for meaningful community participation in NRM [Natural Resource Management]' (Juanwen and Qiu 2006: 187). Of course, the strengthening of links with government comes with tensions and challenges, too. Government structures are notoriously slow to take action and respond to local needs. They have a very chequered history of responding to the needs of the poorest and most vulnerable, and some government structures are positively anti-poor. This is true at international as well as national levels. For example, at the United Nations climate change negotiations, indigenous communities are largely excluded from the formal negotiating sessions because these are conducted on the basis that the nation state is the primary negotiating party.

Top-down approaches to CBA are, to some extent, necessary if adaptation is to be mainstreamed into government policy and planning processes in vulnerable countries, and this will require working with government bodies, which are usually divided by sector. But as discussed, CBA is more about process than outcomes, so this top-down approach needs to be combined with participatory bottom-up processes and an acknowledgement that there are no such sectoral distinctions at the household level. In addition, few governments can legitimately boast strengths in the use of participatory planning processes to help communities identify and respond to their own particular needs.

Integrating effectively

There was a time during the 1990s when very few development professionals were engaged with climate change issues, until it became obvious that climate change was not just an environmental issue. Development agencies such as CARE, ActionAid, Tearfund, World Vision and OXFAM soon realised that climate change was a key driver of poverty for many of the people with whom they worked, and that in some cases their development work could all be in vain if climate change was not incorporated into planning and implementation of project activities. Likewise it soon became obvious that such development organisations were in many cases better placed to help people respond effectively to the impacts of climate change on the ground than climate change scientists or environmentalists. It is now widely accepted that such development agencies have a central role to play in the context of providing support and advice regarding adaptation to climate change.

Bringing an even wider range of professionals, policy-makers and practitioners to the table is important, and in some places already occurring, but evidence from around the world demonstrates that sectoral and disciplinary boundaries still create challenges to holistic integration. While it is widely acknowledged that climate change will affect agriculture, education, health and a range of other sectors and areas of household planning, it is still uncommon to see local responses to climate change being routinely incorporated into planning in all these sectors. Of course, some sectors, planning processes and indeed countries are ahead of others. For example, Huq *et al.* (2003) describe how planning for climate change has been mainstreamed into agricultural policy and practices in Mali, and into national water management policies in Bangladesh, for many years. But elsewhere, other countries and sectors have made much less progress with regards to assessing how climate change will affect them. For example, in his assessment of the extent to which health professionals and national health planning processes in Zambia have started to incorporate climate change impacts, Kasali (2009: 14) notes that Zambia has 'not yet developed any climate-informed policies for the health sector' despite the increased frequency of droughts, floods, extreme heat and shifts in rainy season length experienced over the last decade.

Without this wider understanding of climate change and its impacts, an integrating approach to tackling the problem will be impossible. For CBA to be successfully upscaled, it needs to be incorporated into all planning and implementation of a nation's economic and social development. Even, for example, into the primary and tertiary education provided to the nation's children, because without some understanding of climate change and its risks we cannot expect people to plan ahead for a world which may be very different from that on which their life experiences rest to date.

Being strategic

In an ideal world, CBA initiatives would reach and benefit every single community affected by climate change impacts. But this is not the world we live in, and any upscaling of CBA must therefore be strategic in terms of where efforts are focused. Strategy is important at several levels. At the household level, it might be better to invest in ducks rather than chickens if flooding is becoming increasingly problematic. At the community level, it might make sense to cut losses and relocate a community if sea levels are rising and problems relating to sea water inundation are increasing, rather than investing in strengthening physical barriers to keep the sea at bay. While floating and hanging vegetable gardens have proved useful in coastal Bangladesh in the context of helping communities cope with the ever-increasing problems of waterlogging (Joy 2010), hanging or floating roads or medical centres are not feasible.

Strategic thinking is needed in terms of addressing timeframes. Poor people often focus strongly on immediate needs rather than long-term needs, but adaptation is a long-term process requiring long-term engagement and investment, so engaging a community in a CBA activity might require the provision of 'quick wins' for them until the longer-term benefits of improved resilience or improved adaptive capacity become apparent. Cash payments for communities not to hunt wildlife on their land worked well in certain situations under CBNRM because communities felt the immediate benefit of protecting rather than hunting the animals (Turner 2004).

Lastly, but perhaps most importantly, a strategic approach is needed in terms of assessing how to secure financial resources for CBA activities, where they should be allocated and how. While financial support for climate change adaptation is increasing through multilateral, bilateral and national channels, there will never be enough funding and resources available to reach everyone who is vulnerable. It is therefore important to look for economies of scale when implementing CBA programmes in a certain way, and it might be worth prioritising one area above another if it costs less to replicate activities there. Finance from innovative sources should also be sought, such as public–private partnerships and maybe even vulnerable communities themselves.

Mitigation

Finally, the best way to help vulnerable communities deal with the challenge of climate change is to reduce the scale of the challenge in the first place. As Achim Steiner, Executive Director of the United Nations Environment Programme and Under-secretary General of the United Nations, notes, 'It is clear that the best form of adaptation is mitigation. Emissions of greenhouse gases must be cut and cut fast in order to stabilise concentrations in the atmosphere and stem future temperature rises.'[1]

Additional issues

As noted above, some general guidance for how to scale up CBA can be identified, although they are not without their own special conditions. But there are three additional dimensions that we need to bear in mind for CBA. These are described in more detail below, but in summary can be listed as:

1 scaling down and accessing climate information;
2 reaching the most vulnerable;
3 working with natural systems.

Scaling down climate information

While the focus of this book is on scaling up CBA, it is also important not to forget the need for *scaling down* important information needed for implementing effective CBA. Communities need information on what climate change impacts they will likely be experiencing in their locality in order to make a sound assessment of what risks they will be facing and how to act accordingly. Likewise, local and regional government bodies need climate information at a scale that is relevant for their area of operation if they are to develop appropriate policies and plans to deal with climate change at the local and regional level.

Climate models currently do not provide information at the geographic scale needed to meet these needs. In the case of the small-scale rooibos farmers of the Suide Bokkeveld region in South Africa, local-level observations recorded that climate impacts – in this instance changing rainfall quantities – varied between one valley and the next, and this affected what choices the farmers in these neighbouring areas made with regards to their farming activities (Arendse and Blaauw 2008). The geographic resolution at which climate information is required is therefore often quite small, and yet most existing computer-generated models provide predictions at a resolution that is far too large to be of use.

Likewise, not all climate change models produce the same results, and it can be difficult to know which ones to choose (Ziervogel and Zermoglio 2010). While in parts of the world such as Southern Africa there is general agreement among the models that it will get hotter and dryer, in other areas of the world there is still disagreement between models about which general direction the trends in climate change will move.

The disconnect between what information local people need and what information climate scientists are developing is also true for timescales. Most climate scenarios make predictions on a long-term, multi-year timescale that is of little use to a farmer wanting to know what rainfall he can expect this season and thus what crops he should plant and when. While models often concur on longer-term trends for specific regions, the reality is that model disagreement is often most significant at annual timescales.

Even if accurate within a shorter timescale, however, information generated by models might be of little use to farmers. Predictions of mean annual rainfall, for example, might actually be far less relevant to a farmer than information on when the first rain will come – so he knows when to plant his seeds – or how long the rainy season will last for – so he knows when to harvest.

While models are improving rapidly, and scientists working with these models are becoming increasingly responsive to the fact that the information they produce needs to be relevant to local people, the fact remains that ‘there is a long way to go before climate change scenarios are used effectively in developing adaptation strategies' (Ziervogel and Zermoglio 2010). Wilby *et al.* (2009) estimate that 'Although the use of climate scenarios for impact assessment has grown steadily since the 1990s, uptake of such information for adaptation is lagging by nearly a decade.' Recent research in Africa suggests that farmers often feel climate model data have little relevance for them, which is why they don't use it more widely (Ziervogel and Zermoglio 2010). This is in part due to the geographic and time scales that most current models make information available at, and part of the problem here results from the lack of reliable meteorological data emerging from poor regions like Africa, but it is also due to a lack of awareness on how to find, use and interpret the information out there that does exist. Ziervogel and Zermoglio (2010: 16) stress that

> Bridging the gap between climatologists and end-users will require growth in the people and organizations able to interpret and communicate this information effectively. The translation skills of so-called knowledge brokers are necessary in order to engage a wide range of stakeholders with specific needs.

Reaching the most vulnerable

It is acknowledged that poor people in poor countries will be affected by climate change differently from the wealthy, and that their social status makes them among the most vulnerable in the world to climate change. This vulnerability stems partly from their geographic location in areas more likely to be affected by climate change impacts. For example, low-income squatter settlements in cities are typically located on floodplains or land at the foot of unstable hillsides where wealthier city residents and industry choose not to locate themselves because of the risks. Poor people also have less ability to deal with climate change impacts when they arrive because of their limited financial resources, limited access to skills and technologies and their weak political voice. Poor people are also particularly reliant on climate-sensitive sectors such as agriculture and fishing. They are frequently unable to cope with current climatic (and other) shocks, let alone any future risks related to climate change. It must be acknowledged that the poor have lessons to teach those who have less experience living with the constant threat of loss of livelihood, homestead or life, so there is merit to a discussion about whether they really are more vulnerable just because they are poor. They admittedly demonstrate creativity in survival strategies, but it is important to recognise that these strategies are not typically accompanied by a rise in social status or well-being. Thus, with increasing amounts of funding and other resources being directed at climate change adaptation, it is essential that ways are sought to reach the poorest and most vulnerable sectors of society (Huq and Reid 2007). Reaching the poorest has long been a challenge for the development sector, but the impetus for continuing these efforts in the context of CBA is doubled because it is poor people who have contributed least to the problem of climate change in the first place due to their minimal greenhouse gas emissions.

Providing the governments of poor countries with increases in resources for climate change does not mean it will reach the poor and most vulnerable. Such communities are often marginalised, remote and receive limited services or support from their governments, even when they are able to articulate what support they need. Reaching these hundreds of millions of people will be an immense challenge for any international or national funding mechanisms or policy interventions seeking to support the upscaling of CBA (Reid *et al.* 2009). Ultimately, power structures are at the heart of climate change vulnerability and it will be important to find ways to allow poor vulnerable people to influence policy and be heard in key policy arenas, and ultimately to benefit from any activities initiated to support adaptation if the upscaling of CBA is to be a success.

Working with natural systems

It has already been mentioned that one of the reasons why poor people are particularly vulnerable to climate change is because of their reliance on climate-sensitive natural resource-based sectors such as fishing, farming and forestry. But it is also important to understand the role these natural systems can play in the context of increasing the ability of vulnerable people to adapt. Many CBA initiatives presented at the fifth international conference on CBA stress the importance of natural systems in the context of helping people adapt (Reid *et al.* 2010), and some 56 per cent of Least Developed Country National Adaptation Programme of Action projects reviewed in 2009 had significant natural resource components (Reid *et al.* 2009), but few accommodate the fact that a strong understanding of ecosystems and how they function can also subsequently support community-based adaptation.

Increased use of natural resources, such as collecting wild food sources when crops fail, is a common response to unusual climatic events. It is also not uncommon to learn about case studies supporting the restoration of mangroves with a view to increasing protection from storm surges, or the restoration of forests on steep slopes to protect people from the increased risks of landslides when rainfall is torrential (Reid 2004). The term ‘ecosystem-based adaptation’ (EbA) has been developed to describe the use of biodiversity and ecosystem services as part of an overall adaptation strategy to help people adapt to the adverse effects of climate change (Secretariat of the Convention on Biological Diversity 2009).

EbA, like CBA, is a fairly young concept, but it is increasingly clear that while they have much in common, CBA has much to learn from EbA and vice versa (Girot *et al.* 2012). Ecosystems have limits beyond which they cannot function effectively, and if ecosystems fail then they can no longer play a role in helping communities cope with climate change. Climate change and other pressures are pushing many ecosystems to the limits of their coping capacity, and many, such as coral reefs in certain areas, have passed a tipping point and broken down to the extent that recovery is unlikely (Buddemeier *et al.* 2004).

In many cases we do not yet know exactly how climate change will affect specific ecosystems and what their tipping points are, but we do know that ecosystems are more resilient to climate change if they are in good condition and non-climate stressors, such as habitat destruction, are minimised (Hansen *et al.* 2003). For example, conserving and restoring forests by preventing forest fragmentation, restoring corridors, avoiding over-use of resources and reducing risk of forest fires can increase their resilience to climate change and help to ensure continued community access to natural resources that can ultimately help them adapt to a changing climate. Forests can also help reduce the risk of disasters such as landslides, and forest corridors along gradients can help key species, such as pollinators, seed dispersers and medicinal plants, migrate as temperatures rise and subsequently continue to be available to provide services to local people.

Ecosystems function at different scales, from the local scale, such as a pond, to the large scale, such as an international watershed or river basin. Very often their boundaries do not correspond with social and political boundaries, such as local government administrative units, or village communal land boundaries. CBA may need to take account of these ecosystem boundaries in addition to the more obvious political/social boundaries if adaptation actions based on the ecosystem or ecosystem services are to be successful. For example, if a village sinks a new well to combat drought in its neighbourhood, the effects this will have on the whole aquifer need to be considered if the new well is not to undermine long-term adaptive capacity of both the village and its neighbours. Likewise, the important role that mangroves play in protecting communities from storms as well as providing alternative forms of livelihood (mangroves are excellent fish nurseries and can provide wood for construction and fuel) if crops fail cannot be sustained if plans are not made to allow the mangroves to migrate in response to sea-level rise. In Vietnam and Australia, some communities are setting aside areas behind existing mangroves so they have space to move inland as the sea level rises.

Just as many EbA initiatives could account better for the social complexities that CBA addresses head-on, CBA could also account better for the complexity of ecosystems and ecosystem services to ensure initiatives are more successful. While CBA projects often address the degradation of soils, watersheds, forests, etc., practitioners tend to view adaptation threats and solutions in terms of natural resource silos as opposed to complex ecosystems. This way of thinking can be problematic because, while it focuses on agricultural yields, water, forest products and fisheries, it rarely considers crucial 'second-tier' ecosystem goods and services such as pollination, climate regulation, genetic diversity or the connections between different natural systems.

Conclusions

CBA is clearly a concept that still needs some flesh on its bones. Despite the growing community of people engaged with CBA, there is not complete clarity about where the boundaries of the concept lie. A robust assessment of the success of most actions has yet to be carried out, and there are no real agreed 'standards' for how success or lack of success would be demonstrated.

The baggage brought by lack of rigour in the use of the term 'adaptation' in environment and development fields is also unhelpful.

Yet, some core principles seem to be emerging, such as the need for learning lessons from other similar community-based processes, the need to consider carefully how climate change impacts and information feature in the projects and programmes, and the importance of scale in terms of feasibility and effectiveness.

This volume seeks to provide further fodder for the discussion among scholars, practitioners, donors, community activists and NGOs to enhance our understanding of what CBA is, and develop it into a reliable and successful field. Each chapter in this book attempts to offer insights into different sub-topics, offering material for reflection and upon which to build this body of knowledge and practice.

Note

1 www.100places.com/en/abouttheclimate0/climatechangeandlivelihoods/.

References

Adams, W. and D. Hulme (2001) Conservation and communities. Changing narratives, policies and practices in African conservation. In Hulme, D. and M. Murphree (eds) (2001) *African Wildlife and Livelihoods: The Promise and Performance of Community Conservation*. James Currey Ltd, Oxford.

Adnan, S., A. Barren, S.M. Nurul Alam, and A. Brustinow (1992). *People's Participation: NGOs and the Flood Action Plan*. Research and Advisory Services, Dhaka.

Arendse, A. and R. Blaauw (2008) Tea farming in South Africa. *Tiempo* 68: 16–19.

Ayers, J., M. Alam and S. Huq (2009) Adaptation in Bangladesh. *Tiempo* 72: 9–13.

Buddemeier, R.W., J.A. Kleypas and R.B. Aronson (2004) *Coral Reefs and Global Climate Change: Potential Contributions of Climate Change to Stresses on Coral Reef Ecosystems*. Pew Centre on Global Climate Change, Arlington.

Collins, K. and R. Ison (2009) Jumping off Arnstein's ladder: social learning as a new policy paradigm for climate change adaptation. *Environmental Policy and Governance* 19: 358–373.

Dodman, D. and D. Mitlin (2011) Challenges for community-based adaptation: discovering the potential for transformation. *Journal of International Development*. DOI: 10.1002/jid.1772.

Dumaru, P. (2010) Community-based adaptation: enhancing community adaptive capacity in Druadrua Island, Fiji. *Wiley Interdisciplinary Reviews: Climate Change* 5: 751–763.

Girot, P., C. Ehrhart, J. Oglethorpe, H. Reid, T. Rossing, G. Gambarelli, H. Jeans, E. Barrow, S. Martin, N. Ikkala and J. Phillips (2012) Integrating community and ecosystem-based approaches in climate change adaptation responses. ELAN, unpublished. Available at www.elanadapt.net.

Hansen, L., J.L. Biringer and J. Hoffman (eds) (2003) *Buying Time: A User's Manual to Building Resistance and Resilience to Climate Change in Natural Systems*. World Wildlife Fund, Berlin.

Huq, S. and H. Reid (2007) *Community-Based Adaptation: A Vital Approach to the Threat Climate Change Poses to the Poor*. IIED, London.

Huq, S., A. Rahman, M. Konate, Y. Sokona and H. Reid (2003) *Mainstreaming Adaptation to Climate Change in Least Developed Countries (LDCs)*. IIED, London.

Huq, S., H. Reid and L.A. Murray (2006) *Climate Change and Development Links*. IIED, London.

IIED (1994) *Whose Eden? An Overview of Community Approaches to Wildlife Management. A Report to the Overseas Development Administration of the British Government*. IIED, London.

Joy, K.Z.K. (2010) Hanging vegetable gardens. *Tiempo* 75: 8–11.

Juanwen, J. and S. Qiu (2006) Scaling up community-based natural resource management in Guizhou province, China. In Tyler, S.R. (ed.) (2006) *Communities, Livelihoods and Natural Resources: Action Research and Policy Change in Asia*. Practical Action Publishing/IDRC, London.

Kasali, G. (2009) Adaptation and health in Zambia. *Tiempo* 71: 12–17.

Kepe, T. (1999) The problem of defining 'community': Challenges for the land reform programme in rural South Africa. *Development Southern Africa* 16(3): 415–433.

Magee, T. (2012) *A Field Guide to Community-Based Adaptation*. Routledge, London.

Malla, Y.B., H.R. Neupane and P.J. Branney (2003) Why aren't poor people benefiting more from community forestry? *Journal of Forest and Livelihood* 3(1): 78–92.

Murphree, M.W. (2000) *Community-Based Conservation: Old Ways, New Myths and Enduring Challenges*. Key Address at the Conference on African Wildlife Management in the New Millennium. College of African Wildlife Management, Mweka, Tanzania, 13–15 December.

Pelling, M. (2011) *Adaptation to Climate Change: From Resilience to Transformation*. Routledge, London.

Pretty, J., I. Guijt, J. Thompson and I. Scoones (1995). *Participatory Learning and Action: A Trainer's Guide*. IIED: London.

Reid, H. (2004) Climate change: biodiversity and livelihood impacts. In D. Roe (ed.) *The Millennium Development Goals and Conservation*. IIED, London.

Reid, H., T. Cannon, R. Berger, M. Alam and A. Milligan (eds) (2009) *Community-based Adaptation to Climate Change*. IIED, London.

Reid H., S. Huq and L. Murray (2010) *Community Champions: Adapting to Climate Challenges*. IIED, London.

Schipper, E.L.F. (2009) *Expanding the Community of Community-based Adaptation*. SEI, Stockholm.

Secretariat of the Convention on Biological Diversity (2009) *Connecting Biodiversity and Climate Change Mitigation and Adaptation: Report of the Second Ad hoc Technical Expert Group on Biodiversity and Climate Change*. Secretariat of the Convention on Biological Diversity, Montreal.

Steele, P., N. Fernando and M. Weddikkara (2008) *Poverty Reduction that Works: Experience of Scaling Up Development Success*. Earthscan, London.

Turner, S. (2004) Community-based natural resource management and rural livelihoods. In Fabricius, C., E. Koch, H. Magome and S. Turner (eds) *Rights, Resources and Rural Development: Community-Based Natural Resource Management in Southern Africa*. Earthscan, London.

Uvin, P. and D. Miller (1996) Paths to Scaling-up: alternative strategies for local non-governmental organizations. *Human Organization* 55(3): 344–354.

Wilby, R.L., Troni, J., Biot, Y., Tedd, L., Hewitson, B.C., Smith, D.M. and R.T. Sutton

(2009) A review of climate risk information for adaptation and development planning. *International Journal of Climatology* 29(9): 1193–1215.

World Bank (2005) *Reducing Poverty, Sustaining Growth: Scaling Up Poverty Reduction: A Global Learning Process and Conference in Shanghai, May 25–27, 2004. Case Study Summaries*. World Bank, Washington, DC.

Ziervogel, G. and F. Zermoglio (2010) Informing adaptation. *Tiempo* 75: 12–16.

2 Adaptive capacity*

Rachel Berger, Jonathan Ensor, Kirsty Wilson, Indrani Phukan and Srijita Dasgupta

The IPCC defines adaptive capacity as 'the property of a system to adjust its characteristics or behaviour, in order to expand its coping range under existing climate variability, or future climate conditions' (Brooks and Adger, 2005). Adaptation can be undertaken as a response to a climate-related problem or in preparedness for or anticipation of future climate change impacts. Enhancing the adaptive capacity of local communities to climate change implies an approach to adaptation that is anticipatory. It will be argued in this chapter that adaptation goes beyond reducing vulnerability and preparing for hazards, and involves an ongoing change process where communities can make decisions about their lives and livelihoods in a changing climate. Adaptation is undertaken by various agents, in different ways, at multiple scales. This chapter focuses on what is needed at the local scale to enable poor and vulnerable communities to have the capacity to adapt. A pre-requisite for this is to understand people's and societies' vulnerability and the underlying drivers of this, which extends beyond vulnerability merely to weather and climate factors. Adaptive capacity must also focus on addressing the political, cultural and socio-economic factors that may promote or inhibit individuals and groups from adapting (Smit and Pilifosova, 2001).

In Ensor (2011), adaptive capacity is proposed in terms of the processes which must be in place if communities are to be able to make changes to their lives and livelihoods in response to emerging environmental change. In this view, it is the combination of skills, resources and information that a community can call on and use at a given point in time. Yet these assets – physical, natural, financial and human capital – are themselves only part of the picture. For adaptive capacity, the concern is how networks of relationships define the distribution, access and control of the material and knowledge assets that determine the options that communities can select when looking to respond to environmental change. These relationships are determined by characteristics such as power, culture and gender, and operate across scales, meaning that local communities may need to link or engage with actors and institutions located in different and sometimes distant locations.

The focus of adaptive capacity, then, is on the processes of change, and the character of the formal and informal institutions and networks of relationships that determine the nature of achievable change. These processes must also enable communities to understand and prepare for an uncertain future where the limitations of climate projections are compounded by the complexity of human systems. The challenge of supporting communities' adaptive capacity lies in identifying the mechanisms through which a better-informed image of future scenarios can be generated, assessed and acted on by the poorest in their own best interest. But to be successful, these processes must acknowledge and move beyond the power relationships that are frequently fundamental in sustaining impoverishment and maintaining the unequal distribution of resources and opportunities between different communities.

By thinking of adaptive capacity in this way, the underlying climate change vulnerabilities and risks that a community faces are addressed through the adaptation measures that result from these processes of change. This proposition will be illustrated later in the chapter by examples from programme work by agencies believing that adaptive capacity has been built through their interventions.

Programmatic adaptation approaches can range from addressing drivers of vulnerability (activities that seek to reduce poverty and non-climatic stressors and reduce vulnerability to climate hazards), to building responsive capacity (building robust systems for problem-solving and responding to climate-related stress and shocks), to managing climate risks (incorporating climate information into decision-making), and finally to confronting climate change itself (addressing impacts associated with climate change) (McGray *et al.* 2007).

Different framings of adaptive capacity

The concept of adaptive capacity has received increasing attention in recent years as academic and action-oriented research programmes have assessed what it means in practice (Jones *et al.* 2010). For example, a consortium of organisations led by Christian Aid, Plan International, and the Institute of Development Studies, has developed, and is currently testing, an approach incorporating adaptive capacity called Climate Smart Disaster Risk Management (CSDRM; Mitchell *et al.* 2010).

The CSDRM approach has made promising advances in operationalising, monitoring and evaluating adaptive capacity, and has done so by linking to disaster risk reduction (DRR), climate change and development policies and programmes. CSDRM also identifies a set of indicators that can be tailored and contextualised to, for example, community-based adaptation (CBA), in order to assess progress towards integration. It offers a set of action points, guidance and indicators that link together uncertainty, poverty, vulnerability and adaptive capacity. Practitioners and policy-makers are using this approach to drive towards long-term, pro-poor disaster risk management (DRM). The CSDRM approach draws on the extensive participation of over 500 practitioners, policy-makers, scientists and academics working on climate change, disasters and development communities in 'at-risk' countries in Africa and Asia. In this chapter, two case studies use this approach, which comprises three 'pillars' of action. These address:

1 Uncertainty – to improve information on how hazards are changing with detailed risk assessments from diverse sources of knowledge (e.g. local, climatology, social science).
2 Adaptive capacity – to enable institutions and networks to develop new skills, knowledge and resources needed to enhance capacity to adapt to climate change.
3 Poverty and vulnerability – to support communities to build their resilience by addressing how disasters and climate change can increase existing inequality, lack of resource access and unsafe conditions.

Under pillar two, enhancing adaptive capacity is defined in terms of:

- strengthening the ability of people, organisations and networks to experiment and innovate through encouraging flexible and effective institutions;
- promoting regular learning and reflection to improve the implementation of policies and practices;
- ensuring policies and practices to tackle changing disaster risk are flexible, integrated across sectors and scale and have regular feedback loops;
- using tools and methods to plan for uncertainty and unexpected events.

Two programmes in India, described in the next section, show how adopting the CSDRM lens can help strengthen integration between disasters, climate change and poverty/vulnerability and in doing so also promote adaptive capacity.

The Africa Climate Change Resilience Alliance (ACCRA) consortium[1] take an alternative approach to adaptive capacity, identifying five characteristics in its Local Adaptive Capacity Framework (LACF) (Jones *et al.* 2010):

1 *The asset base*: the various financial, physical, natural, social, political and human capitals necessary to best prepare a system to respond to a changing climate. This category incorporates the importance of social networks.
2 *Institutions and entitlements*: the ability of systems to ensure equitable access and entitlement to key resources and assets for all groups, including the most marginal and vulnerable to climate change. Access to key resources, participation in the decision-making process and empowerment are key elements of the characteristic.
3 *Knowledge and information*: successful adaptation requires information and understanding of future change, knowledge around adaptation options, the ability to assess them, and the capacity to implement the most suitable interventions. Systems must be in place to distribute relevant information at both national and region scales and forums for dialogue and discussion among all stakeholders must exist.

4 *Innovation*: adaptive capacity relates to the system's ability to support innovation and risk taking. Innovation can be planned, high-tech oriented and geared towards large-scale innovations; or it can be autonomous, local-level initiatives that help innovate or adapt to changes to the local climate. An enabling environment that promotes and allows for experimentation and the exploration of niche solutions is required, which can also protect against risks of failure associated with innovation.
5 *Flexible forward-thinking decision-making and governance*: ensuring that local organisations are informed on future climate impacts and take appropriate measures to plan for the future. Flexibility for systems, and the institutions that govern them, to evolve and adapt to a changing environment is crucial for adaptive capacity.

Use of the ACCRA framework to analyse a project in India is described in the next section. The view of adaptive capacity developed by Practical Action offers a somewhat different approach, looking to identify how development actions can *support* adaptive capacity rather than attempting to locate the components of adaptive capacity (Ensor, 2011). Developed from an understanding of resilience and complexity, this approach suggests that adaptive capacity can be best supported through a focus on three critical dimensions of institutions and relationships:

1 The *power sharing* arrangements that are in place to expand communities' voice and influence over decision-making.
2 The sources of and processes that give rise to the *knowledge and information* that inform adaptation decisions.
3 The availability of *experimentation and testing* of adaptation options that are relevant at the local level.

'Power sharing' draws attention to the processes through which development actions proceed, requiring support for new forms of participation that seek to integrate the views of diverse stakeholders from different scales and with different power into shared decision-making. Second, 'knowledge and information' are necessary for effective decision-making that integrates the challenges and opportunities likely to be faced in the future. Communities will need to bring new sources of information into their current knowledge base and decision-making framework in order to address the emerging challenges of climate change, combining knowledge gained from different worldviews and experiences. 'Experimentation and testing', finally, are necessary to inform decision-making processes through contextualised (location-specific and culturally relevant) understandings of proposed responses to the challenges of climate change. As Figure 2.1 suggests, when these three components are supported through development actions, they link together to form a learning system in which communities gain the power, knowledge and opportunities to form and revise their responses to climate change. The result is a framework of support for adaptive capacity that enables communities to build their resilience. Reflecting

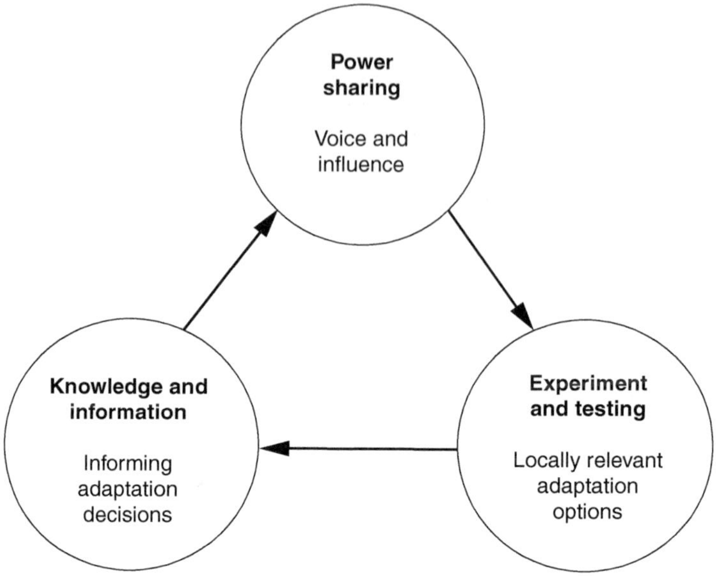

Figure 2.1 A flexible learning system approach to supporting adaptive capacity (source: adapted from Ensor, 2011).

the attributes of adaptive capacity identified by ACCRA and CSDRM, the emphasis is on ensuring a flexible, knowledge-based learning system that enables innovation and testing through forms of participation that recognise and build on the entitlements of the poorest.

Adaptive capacity in practice

The following case studies provide some initial insights into how development projects can support adaptive capacity. The conclusion reflects on these examples in terms of the different dimensions of development support for adaptive capacity identified in Figure 2.1.

The Western Orissa Rural Livelihoods Programme

This section describes the Western Orissa Rural Livelihoods Programme (WORLP), a government of Orissa initiative. WORLP is managed by the Orissa Watershed Development Mission, and is a joint venture of the government of Orissa and the UK Department for International Development (DFID), with technical assistance to the project provided by NR International of the UK. The CSDRM framework is used here to examine how WORLP has altered the processes of disaster planning and response and thereby impacted on the adaptive capacity of local communities.

Orissa State Disaster Management Authority (OSDMA) is at the centre of state policy-making and has an operational mandate to cover management of disaster relief and reconstruction. While OSDMA coordinates disaster risk management activities centrally, district administrations initiate disaster management mechanisms through committees at different levels. Village-level disaster management planning is carried out by Panchayat Disaster Management Committees. Village headmen are instructed to initiate the constitution of the village disaster management planning committees, while community members and NGO volunteers support the development of community contingency plans. In this way local knowledge is integrated into local planning, although indigenous knowledge about weather forecasting has not been systematically integrated into all planning processes.

The creation of DRM institutions in India has drawn momentum from disaster events, starting after the Orissa cyclone and accelerating following the 2004 South Asian tsunami. A Disaster Management Bill was tabled in Parliament in 2005 and enacted structures at all levels – national, state, district and block – to prepare for and reduce the effects of disasters. At the state level, OSDMA is an autonomous nodal agency and has played a key role in coordinating with various line ministries to integrate DRM into ongoing flagship programmes. That said, DRM is not yet mainstreamed into all development projects.

OSDMA also had the space to create stronger institutional coordination with NGOs, developing its own database on NGO initiatives and holding regular consultations. It has established emergency coordination among government departments, state headquarters and district administration, and with NGOs; it has ensured a quick and efficient response to floods. OSDMA has strengthened community disaster preparedness, through mock drills, training, contingency planning and the formation of village taskforces, all of which have enabled people to cope better with floods. Close collaboration with NGOs has enabled OSDMA to assess capacity and training needs, prepare action plans and help match government funding to NGO rehabilitation activities in cyclone-hit areas.

The institutional set-up of WORLP has allowed a high level of autonomy and flexibility. Activities focused on increasing people's ability to adapt to and cope with climate-related stress have been implemented in a quick, effective and participatory way, through a direct chain of command. WORLP has achieved meaningful programmatic links and training initiatives across natural resources management, livelihoods, micro-enterprise and other related developmental programmes. A critical capacity-building team has been established in the district Project Director's office.

A four-member team specialises in livelihoods, microenterprise, natural resource management and monitoring and evaluation. A three-member livelihood support team comprising specialists from agriculture, microenterprise and social development sectors supports the implementation and monitoring of activities on the front-line (known as 'watershed plus'), in conjunction with 'watershed development' teams of village volunteers. Through these interventions almost 6,000 self-help groups have been trained to cultivate local action.

Non-refundable grants, either for consumption or assets, have been provided to those households deemed by the community to be 'very poor', and a revolving fund provides loans for microenterprises.

What is not clear is the extent to which the capacity of the local community has been built at household level, rather than the capacity of local NGOs. It is unclear the extent to which the local community, and the associated institutions, will, as a result of government-led initiatives (WORLP and OSDMA), be better able to deal with future uncertainty, which may lead not just to more frequent shocks, but also to slower onset disasters.

CSDRM in enhancing adaptive capacities and livelihood resilience for small and marginal farmers in flood affected areas of Uttar Pradesh, India

This example describes a case study using the CSDRM lens to analyse an existing project in Uttar Pradesh, India. The programme focuses on agriculture, in a context where agricultural production comprises 34.8 per cent of state income and employs 66 per cent of total workers (or 35.5 million people) and 90 per cent of the population are small-scale farmers or landless people. Rainfall patterns and volume have changed, with more flooding and dry spells at unpredictable times. The impact of this has been reduced agricultural production, migration in search of work and decreasing biodiversity.

In response, the project has sought to combine local and scientific knowledge on changing climate-related hazards to encourage adaptive agricultural practices, agricultural models and farm sub-systems that are better able to withstand the impacts of the changing hazards. These measures were supported by initiatives to enable farmers to improve the management of their land and crop cycle. A time and space management system was also developed and widely adopted by small-holder farmers, which alters the timings of the conventional cropping cycles to ensure higher returns in the context of a hazard-prone and changing climate. Time-management techniques include sowing early so that crops are harvested before flooding, to reduce the risk of loss from a specific hazard. Techniques also include water-resistant crops, which can grow even if a field is inundated with floodwater, and sowing when water has receded from the field post-monsoon during rehabilitation and recovery.

Looking at this work from the CSDRM perspective, it can be seen that emphasis has been placed on strengthening understanding of emerging risks from climate change and ensuring access to climate information for farming work. This has been combined with elements that build adaptive capacity, such as supporting innovation and experimentation. For example, the project supports the new management practices outlined above, as well as using permaculture to build resilience (for example see Borron, 2006).

Another key element for adaptive capacity is the promotion of regular learning and reflection. This is incorporated where farmers are supported in their needs for information about new farming technologies through farmer field schools. This is combined with support for the formation of farmers' self-help groups and strengthening links with the government departments and formal extension system.

Strengthened learning and enhanced adaptive capacity is integrated strongly with DRM and dealing with uncertainty through village resource and information centres that have been established. Specific disaster mitigation groups have been set up to act when an extreme climate event is forecast.

Reflecting on the Uttar Pradesh programme described above, the CSDRM analysis has concluded that integration of the activities of different stakeholders working to manage disaster risk and uncertainties with work on adaptation and development is key. In addition, it is crucial to strengthen the ability of community-based groups to experiment, innovate, access appropriate technology and utilise their local knowledge effectively.

The extent to which power relations between community members and local institutions has changed as a result of the programme on addressing poverty, vulnerability and its causes is not clear and probably cannot be assessed so soon after the intervention. That said, it has been noted that self-help groups encouraged the involvement of women in savings and credit schemes, and the practice of food production with little or no use of fertilisers, pesticides or external inputs to food production, thereby incorporating perspectives on gender inclusion.

Africa Climate Change Resilience Alliance

The ACCRA programme aims to produce a number of case studies of projects implemented by the consortium members to examine the extent to which projects not designed as adaptation projects succeed in building adaptive capacity, and to suggest ways in which project design could be adjusted to do so. Wilson and Getnet (2011) explore the findings from a case study produced under ACCRA by Haramaya University in Kaseja kebele, a lowland agricultural area in the West Hararghe zone of Ethiopia, where CARE is implementing the Household Asset Building and Rural Empowerment for Transformation (HIBRET) project. The paper highlights major aspects of the LACF which are addressed by CARE's work, as well as those which remain overlooked in this approach to reducing vulnerability. The project promoted supplementary livelihoods activities to support household asset building, including vegetable production and the introduction of new varieties of haricot bean and maize. However, inadequate skills training and poor market access for surplus production have raised the risk that the activities could actually be mal-adaptive: livelihoods have not been made more resilient, and land and irrigation water has been used that could instead have been devoted to more important food or cash crops.

The five characteristics described under ACCRA's LACF are a starting point to conceptualise adaptive capacity at the local level, and an entry point for discussion of how to increase community- or household-level adaptive capacity. Using this lens to review the case of Kaseja highlights a number of possible recommendations to guide the development of programmes in ways that might more effectively build adaptive capacity (Wilson and Getnet 2011). Investment in soil and water conservation activities was a central part of the project, but the case study indicates a need for a greater focus on the institutions governing access to scarce communal resources and a need for longer-term planning about how these will be managed. The importance of institutions, such as markets, and access to them, is highlighted, as is the need for infrastructure improvements to improve physical access as well as access to traders' networks. Lack of technical knowledge, despite the project's efforts, is still limiting adaptive capacity.

Finally, this project highlights a challenge facing adaptation programmes in marginal areas for food production. In Kaseja, only two people had ever gone to secondary school. The ACCRA analysis suggests that education and job creation are needed to open up opportunities for livelihoods that are less dependent on climate-sensitive natural resources, otherwise efforts to strengthen adaptive capacity will continue to be limited in their impact.

Q'emikuspa: adaptation measures to protect and improve the livelihoods of alpaqueros communities

A three-year project funded by Oxfam in the Peruvian Andes near Cusco is being implemented by Practical Action and Asociacion Proyeccion, a local organisation with good knowledge of the characteristics of high-altitude communities, and a track record in community organising, training and mobilisation. The Q'emikuspa comprises community-level work and research to validate technologies and processes. This project is working with 22 communities at altitudes between 3,500 and 5,000 m above sea level. The main livelihood activities at this altitude are livestock and alpaca raising and small-scale agriculture for household consumption. The communities that have inhabited the Andes over 10,000 years have developed technologies for agriculture and water management, such as terracing, irrigation and soil improvement. They have also developed techniques to mitigate effects of climate, use of micro-environments with a diversity of species, prediction systems that rely on indicators from the environment and astronomy, processing of crops (dried potato, chuño,[2] etc.), pest and disease control through crop rotation, which are still actively used today as coping strategies for the extreme climate.

The project is enabling innovation in water management and agricultural production, and access to information for early warning of extreme weather events. To ensure local ownership and adequate institutional capacity, formal agreements with local authorities (provincial and district level) and technical and education institutions have been entered into; community civil defence committees have been formally recognised by local government so that they can liaise with the official district and regional civil defence committees. The project specifically focused on improving women's positions in their households and communities, since women are disadvantaged by local culture and affected by male migration and domestic violence. National legislation in Peru requires participatory budgeting at the district level, but engagement demands a high level of access to information and understanding of the process. The project has sought to support communities in engaging in this process.

In the two years that the project has been implemented, some important results have emerged which are helping to build the community's adaptive capacity as well as strengthen its relationships with important local actors. In keeping with the Practical Action framework outlined above, the project has brought together indigenous peoples and local authorities (district, provincial, sub-national). This is an important stage in building consensus among stakeholders and equalising power between them. Information and knowledge are required for this process. As a result of improvements in access to information and training, communities can calculate the intensity of rainfall in the area, access information on weather and take action to protect their animals, children and elderly relatives. In this project, the process of engaging local communities does seem to have built adaptive capacity to cope with future climate-related challenges.

The importance of an enabling policy framework

There has been little NGO programme or project work so far that has focused on developing an enabling policy framework. Generally, most international NGO programmes contain only a small component of advocacy and policy-influencing, for two reasons: (1) it is not often included in the components that a donor will fund, and (2) the skills required are not often in abundance in programme partners' or international NGO country office staff. A case study conducted in Cox's Bazaar in Bangladesh highlights how important enabling policies are to enabling adaption, and hence to the building of adaptive capacity (Dasgupta 2011).

Dasgupta conducted a rapid roving survey in late 2010 in five Upazilas including Cox's Bazaar, comprising traditional fishing communities of the coastal belts of Bangladesh, who belong to the Hindu caste of the *Jaladas* (slaves of the sea). In the 1960s, during the green revolution when there was an increase in the cultivation of rice, there was a notable decrease in inland fishing as more land came under rice cultivation. As a result of this, inland fisher families migrated to the coastal belts to earn extra income. These incoming communities bought (and continue to buy) the fish caught by the traditional fishers and market it in a manner that is leading to over-exploitation and unfair allocation of resources. The traditional fishers also have to contend with indiscriminate fishing by other groups.

The traditional fishermen learned their fishing skills from their ancestors and only have the knowhow to catch fish in the bay, a limitation that has led to livelihood insecurity. They notice the decline in their catch compared with the past, but they have no access to means of learning other skills. This group is highly marginalised: one fisherman pointed out that the government does not even have a record of the number of fishers' households in Bangladesh. To reduce this community's marginalisation and support them in adaptation would require policies to ban overfishing, as well as providing resources to enable fishermen to access modern fishing equipment, and also market access and regulation to ensure a fair return. Training in risk reduction initiatives and early warning of storms and cyclones is also needed. However, the reality that fish stocks are declining and that the number of active fishermen needs to be reduced accordingly must also be addressed. Since power relations have been one root cause of this community's declining livelihood (as a result of more powerful incomers and outsiders competing for the fish catch), combined with marginalisation in terms of lack of government resources reaching them, national government commitment and action will be required to help redress this situation.

Conclusion

Mitchell *et al.*'s (2010) report on CSDRM emphasises the need for donors and governments to support flexibility and innovation through joined-up, strategic and collaborative approaches, concluding that:

- Adaptive capacity is central to improving ways of working and will require systematic investment in skills and innovation.
- Rights and access to services provide the foundation on which DRM and adaptive capacity can be promoted.
- Dealing with changing risk and uncertainty requires new knowledge that can be blended and brokered in a way that aids effective implementation.

An extra dimension to these conclusions is offered by ACCRA and Practical Action's focus on the dynamic processes central to adaptive capacity. Both draw attention to the power relations that may determine adaptive capacity at the local level – and thus to the wider policy framework that (as Dasgupta suggests) circumscribe local rights and entitlements (Jones *et al.* 2010; Ensor 2011).

The ACCRA consortium recognises that since the precise nature of climate impacts are impossible to predict with the level of certainty required for definitive policy decisions, adaptation must manage uncertainty and foster adaptive capacity. Adaptation is seen therefore not as a choice between reducing general vulnerability or preparing for specific hazards, such as floods, but a process of ongoing change whereby people can make informed decisions about their lives and livelihoods in a changing climate. Learning and being able to adapt is as important as any specific adaptation intervention (Pettengell 2010). There is much in common here with the framework set out by Practical Action (Ensor 2011) and the elements with CSDRM. However, to quote one of CSDRM's consultation workshop participants (Mitchell *et al.* 2010: 37): 'The DRM community hasn't been good at looking at long term timeframes or capacities to enhance people's ability to progress over the longer term; this [CSDRM] approach encourages us to do this.'

Recognising the need to adopt a longer timeframe and address an inherently uncertain future is a key difference between undertaking 'adaptation' (addressing a given set of changed circumstances), and 'building adaptive capacity' (enabling people to live in an ever-changing and unpredictable future). The examples in this chapter provide some insight into how this can be done, as summarised in Table 2.1.

Table 2.1 Practical examples of support for adaptive capacity drawn from the four case studies

Project/programme	*Western Orissa Rural Livelihoods Programme (WORLP), India (Christian Aid)*	*Uttar Pradesh, India (Christian Aid)*	*Household Asset Building and Rural Empowerment for Transformation (HIBRET), Ethiopia (implemented by CARE, case study by ACCRA)*	*Q'iemikuspa, Peru (Oxfam/ Practical Action)*
Power sharing	• Communities involved in development of local plans • Relationships built between government and NGOs • Flexible institutional arrangements	• Integration of stakeholders and strengthened links between farmers' groups, government departments and extension	• Identified need for institutions to govern access to scarce communal resources	• Brought together indigenous peoples and local authorities to build consensus and equalise power relationships • Formal agreements with local authorities • Raising women's voices
Experimentation and testing	• New livelihood opportunities	• Support for innovation and experimentation with farming strategies • Opportunities for reflection and learning	• Opportunities to experiment with alternative livelihood practices	• Innovation in water management
Knowledge and information	• Training in natural resource management, livelihoods and micro-enterprise	• Combining local and scientific knowledge • Strengthening understanding of emerging risks • Training in new agricultural practices	• Need for skills training and access to locally relevant knowledge that draws on people's understanding of their environment	• Access to information on extreme weather events • Capacity building on government participatory budget process • Information and knowledge to build consensus

Scaling up

For CBA to go to scale involves programmes being implemented by many agencies other than international NGOs working with local partners. It will involve local and national governments, and communities themselves, with support, taking charge of their development. A framework that identifies clearly the elements of adaptive capacity will assist with the training and capacity building needed to go to scale. The three frameworks examined in this chapter all have common elements, and if applied widely will contribute to the upscaling of CBA.

At the same time, this chapter has argued that supporting adaptive capacity may require a long-term approach with multiple phases of engagement, gradually building both access to adaptation measures and the necessary assets and resources to implement those measures. As development professionals, we are currently, therefore, in a situation where it is too early to assess whether adaptive capacity really has been built, and are perhaps only just at the stage of being able to identify indicators appropriate for measuring it (Rossing *et al.*, this volume). It is clear from the preference for donors to support pilot programmes on adaptation (see the PPCR, Pilot Programme on Climate Resilience) and research programmes (IDRC Climate Change Adaptation in Africa) that they do not consider enough is understood about the nature of adaptation to climate change for scaling up to be appropriate at this stage. There is, then, work for NGOs and researchers to do in showing how the different frameworks for adaptive capacity clearly show the nature of the adaptation process, and offer a methodological approach to scaling up.

Notes

* This chapter draws on case studies prepared by Anwara Begum (Caritas), Krishna Lamsal (Li-Bird), George Kasali, John Ajigo, Lorena del Carpio (Oxfam LA).

1 ACCRA is a consortium made up of Oxfam GB, the Overseas Development Institute (ODI), Save the Children Alliance, Care International and World Vision International. The programme is funded by the UK's Department for International Development.

2 Chuño is a freeze-dried potato product traditionally made by Quechua and Aymara communities of Bolivia and Peru.

References

Borron, S. (2006) *Building Resilience for an Unpredictable Future: How Organic Agriculture Can Help Farmers Adapt to Climate Change*, FAO, Rome.

Brooks, N. and Adger, W.N. (2005) *Assessing and Enhancing Adaptive Capacity*, UNDP, New York.

Dasgupta, S. (2011) 'Adaptation capacity of fishing communities to climate vulnerabilities and changes for improving their livelihoods: a case study conducted in Cox's Bazaar District of Bangladesh' Presentation at the Fifth International Conference on Community Based Adaptation and additional unpublished material.

Ensor, J. (2011) *Uncertain Futures: Adapting Development to a Changing Climate*, Practical Action Publishing, Rugby.

Jones, L., Ludi, E. and Levine, S. (2010) *Towards a Characterisation of Adaptive Capacity: A Framework for Analysing Adaptive Capacity at the Local Level*, ODI, London.

McGray, H., Hammill, A., Bradley, R., Schipper, E.L. and Parry J.-E. (2007) *Weathering the Storm: Options for Framing Adaptation and Development*, Washington, DC: World Resources Institute.

Mitchell, T., Ibrahim, M., Harris, K., Hedger, M., Polack, E., Ahmed, A., Hall, N., Hawrylyshyn, K., Nightingale, K., Onyango, M., Adow, M. and Sajjad Mohammed, S. (2010) *Climate Smart Disaster Risk Management*, IDS, Brighton.

Pettengell, C. (2010) *Climate Change Adaptation: Enabling People Living in Poverty to Adapt*, Oxfam, Oxford.

Smit, I.B. and Pilifosova, O. (2001) 'Adaptation to climate change in the context of sustainable development and equity', in *The Third Assessment Report of the Intergovernmental Panel on Climate Change*, Cambridge University Press, Cambridge, pp. 879–912.

Wilson, K. and Getnet, M. (2011) *Investigating How Development Interventions Increase Community-Level Adaptive Capacity in Ethiopia*, n.p.: Alliance for Climate Change Resilience in Africa.

3 Adapting development

How wider development interventions can support adaptive capacity at the community level

Eva Ludi, Sarah Wiggins, Lindsey Jones, Jo Lofthouse and Simon Levine

For most developing countries, climate change adds another layer of complexity to already existing development challenges such as high levels of poverty and inequality, rapid population growth, underdeveloped markets, poor infrastructure and service provision and weak governance systems. Modelled climate projections suggest the impacts are likely to exacerbate these developmental challenges, 'as more floods, more droughts, more strong storms, and more heat waves' make development policy and practice more complicated (World Bank 2010: 1). Poor people are the most vulnerable to the impacts of climate change, both because of their existing vulnerability, and because of their dependence on weather-dependent livelihoods.

Current climate models agree that there will be change, and that it will vary greatly from place to place. However, it is rarely possible to know for certain exactly what that change will be at the local level. Climate change affects people directly (e.g. through changing rainfall patterns, increasing temperatures, etc.) and indirectly, by exacerbating other changes, from global and local crop prices, migration patterns, possible conflicts over dwindling natural resources, etc. These, too, are constantly changing, magnifying the uncertainty of the effects of climate change (Ludi *et al.* 2012).

Climate change has implications for the design and implementation of development interventions. Progress towards reducing poverty is very likely to be undermined by climate change: the OECD estimates that up to 50 per cent of development assistance may be at risk because of climate change (Agrawala 2005). People and places where levels of development are low are most at risk from a changing climate (Smith *et al.* 2003). In many developing countries, those most likely to be vulnerable to climate change impacts are the marginalised and those who lack the opportunities, resources and support to adapt: notably the poor, elderly and the young, and, in many societies, women and ethnic minorities. Causes of vulnerability relate both to the assets people own and to the accountability and effectiveness of institutions and, for many population groups, long-term processes of social, economic and political marginalisation.

However, simply 'adding on' mitigation and adaptive strategies to unsustainable development pathways will not work (Parry 2009). Similarly, repackaging existing livelihoods or disaster risk reduction programmes as climate change adaptation work, often conceptualised as community-based adaptation projects (i.e. operating at local level in vulnerable communities, implementing activities aimed at strengthening the capacity of local people to adapt to climate change, using participatory processes involving local stakeholders to identify interventions (e.g. Ayers and Forsyth 2009, cited in Dodman and Mitlin 2010), presents the risk that climate change is not adequately taken into account when programmes are designed. As climate change is but one of many changes people face, just supporting them to adapt to negative impacts of climate change without addressing underlying vulnerabilities would miss the point altogether. People and communities need to be supported to adapt to changing circumstances overall. Supporting local adaptive capacity, therefore, cannot just be packaged as a 'community-based adaptation' (CBA) project with a limited time frame, scale and scope, but needs to be an intrinsic part of any development intervention, and of attempts to scale-up and broaden localised CBA interventions.

We posit that the contributions of development interventions, if done correctly, can also enhance the adaptive capacity of people and communities to deal with negative impacts of climate change. Building on ACCRA's[1] and Tearfund's research and experience, we further conclude that it does not matter so much under what sector (or approach) an intervention is being carried out, but much more *how* and with which perspective. It doesn't matter tremendously whether interventions are labelled as 'livelihoods support and diversification', as 'social protection' or as 'disaster risk management', what matters is how development or adaptation interventions are done – whether they are supporting adaptive capacity or not – and that means whether they take into consideration in their activities the different characteristics of adaptive capacity and are able to move beyond the local scale at which many development projects operate, to address the root causes of vulnerability and marginalisation.

This chapter builds on research carried out by the ACCRA Consortium[2] between 2009 and 2011in eight study sites in Ethiopia, Uganda and Mozambique (see Ludi *et al.* 2011 for Ethiopia; Jones *et al.* 2011 for Uganda; and Arnall 2012 for Mozambique) and by Tearfund.[3] In the ACCRA study sites (Tables 3.1–3.3) one of the consortium members implemented development interventions. The different sites were chosen to represent different livelihoods, agro-ecological characteristics and types of project intervention, including disaster risk reduction (DRR), social protection and livelihoods support programmes, in reaction to the different development challenges and climate hazards each site faces. The next section summarises the current discussion on the development–adaptation interface and briefly outlines different frameworks developed or currently under development that aim at conceptualising the adaptation–development interface with the aim to support adaptive capacity at scale. The third section discusses findings resulting from ACCRA's research and experiences from Tearfund's projects in terms of where development interventions fail or succeed to support adaptive capacity and why. It consciously does not use case studies to highlight the points made but presents a synthesis of findings from the ACCRA research

Table 3.1 ACCRA research sites in Uganda (source: Levine *et al.* 2011)

Site	*Bundibugyo*	*Gulu*	*Kotido*
Geographical zone	Highland and lowland	Woodland	Semi-arid
Principal climate hazard	Floods	Poor rains, variable rainfall	Poor rains, variable rainfall
Main livelihood source	Rain-fed agriculture for market; cash crops important in lowland areas	Rain-fed agriculture for market	Pastoralism; subsistence rain-fed agriculture
Development interventions	*Rwenzori Livelihoods and Disaster Preparedness Support programme* • Empowerment of poor people to achieve sustainable livelihoods, influence those with power over them and ultimately improve their standard of living • Support community priorities such as improving livelihood diversity, food security, agro-processing activities and DRR planning	*'Roco Kwo' programme* • Multi-sectoral initiative addressing sustainable livelihoods, peacebuilding, conflict resolution, gender equity, psychosocial support and gender-based violence • Agricultural input provision • Establishment of village savings and loan schemes	*North Karamoja Pastoral Development programme* • Support resilient pastoral livelihoods • Working with local government to empower rural communities • Supporting community-based groups

and Tearfund's experience to highlight their wider importance and implication beyond the specific project experiences. These lessons learned, finally, are pulled together in the fourth section, which discusses options for integrating adaptation and development interventions in view of scaling up project interventions so they contribute to adaptive capacity and sustainability.

The development–adaptation interface

Two aspects of the development–adaptation interface are crucial as they also relate to CBA: first, how development approaches can contribute to reduced poverty in the short term while at the same time help enhance communities' capacity to adapt to a changing climate over the longer term. Second, how to re-focus and change the way development interventions are done to actively support adaptive capacity at scale and prevent actions that undermine adaptive capacity. Although many CBA interventions claim to link adaptation and development, too often such

Table 3.2 ACCRA research sites in Ethiopia (source: Levine *et al.* 2011)

Site	*Ander Kello*	*Kase-hija*	*Wokin*
Region	Afar	Oromia	Amhara
Geographical zone	Lowland (dry *Kolla*)	Lowland (*Kolla*)	Highland (*Dega*)
Principal climate hazard	Cyclical drought	Drought, floods	Erratic rainfall, floods, hailstorms
Main source of livelihood	Pastoralism, increasing agro-pastoralism	Mixed crop cultivation, *Khat* is an important cash crop	Mixed crop cultivation, livestock, increasingly market-oriented
Key programme interventions	*PILLAR* – Preparedness Improves Livelihood Resilience • Improve drought preparedness through protecting and diversifying the livelihood assets of pastoralists in drought-prone areas	*HIBRET* – Household Asset Building and Rural Empowerment for Transformation, including PSNP • Reduce food insecurity and increase community resilience by (1) protecting household assets and community resources; and (2) increasing agricultural output through integrated natural resource management and strengthened civil society	*Agricultural Scale Up Programme* • Policy influencing, enhancing market access and improving gender equity
Focus of intervention	DRR	Social protection and sustainable livelihoods	Sustainable livelihoods

projects are simply re-packaging standard rural development projects and thereby completely missing a focus on supporting adaptive capacity of people to deal with the many changes they face over the long term, while other CBA projects entirely focus on addressing actual or potential impacts of climate change without considering underlying vulnerability, which is the root cause of people's inability to deal with change. Therefore, what is needed is a framework for adaptive capacity that allows addressing these links between adaptation and development and that is not just re-packaging or re-labelling development but that goes beyond simply achieving measurable impacts on poverty but actually supports people in building their ability to deal with change, whether weather-related or other changes.

Linking adaptation and development more closely is not easy, and is made particularly difficult because of the long timescale over which many aspects of

Table 3.3 ACCRA research sites in Mozambique (source: *Levine et al.* 2011)

Site	*Chibuto*	*Caia*
Region	South (Gaza Province) Limpopo River Basin	Centre (Sofala Province) Zambezi River Basin
Principal climate hazard	Drought	Floods
Main source of livelihood	Agriculture for subsistence crop production; cattle-rearing; migrant labour	Agriculture for subsistence crop production and cash income; cash crops; casual farm labour
Key programme interventions	Integrated rural development programme • Food and seed distribution • Micro-credit • Livelihood diversification	Livelihood diversification • Formation of groups and associations

climate change will unfold – and hence climate change adaptation will have to evolve. Many projects – whether CBA projects or more general rural development projects – will find it difficult to evaluate the success of their adaptation interventions on development outcomes in the face of rapidly evolving stresses and risks that are changing the contexts in which development takes place ('shifting baseline') (Brooks *et al.* 2011).

Many interventions aimed at supporting adaptation at the local level face three key challenges: (1) the short timescale that characterises small projects, which is inadequate to achieve the required behavioural changes; (2) the community focus, which does not take into account the actual drivers of vulnerability and often conceptualises 'community' as a homogeneous group and thus fails to take into account inequality and power relations both at the local level and between local, national and international levels; and (3) the limited scope, focusing on just one change – climate change – although people are faced by a whole range of changes (see also Dodman and Mitlin (2010), who come to similar conclusions). Community-based projects can be successful in supporting people to improve their livelihoods and well-being, but often only as long as projects are continuing to provide support. More often than not, such community-focused project interventions fail to address issues that undermine sustainability, that would support individuals, households and communities to develop the required capacities that will allow them to adapt to future changes. They also often don't address drivers of vulnerability, including pervasive power relations at multiple levels, of which many lie beyond the local level.

Different authors have attempted to develop practical frameworks that attempt to bring both development and adaptation together. The Local Adaptive Capacity (LAC) framework (Figure 3.1) developed for ACCRA, which, building on the fact that development implies the ability to adapt to a constantly changing context, starts with assessing what people and communities can do for

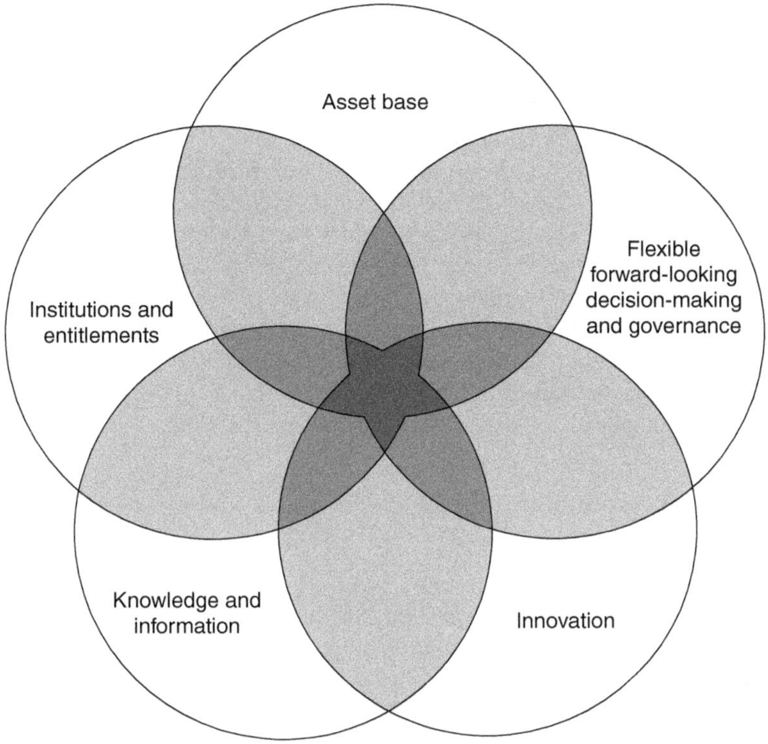

Figure 3.1 The ACCRA framework for thinking about LAC (source: Jones *et al.* 2010).

themselves to maintain or improve their well-being, and the context in which they make their living.

Livelihood strategies and livelihood outcomes do not only depend on assets people have, but also on the institutional environment that governs people's entitlements to assets and influences the use to which people can put these assets and what they can get out of them. The institutional environment also influences people's interactions, their innovative behaviour, how information is used and how decisions, both short term and long term, are made. In short, people's and communities' adaptive capacity – whether to weather-related or other change – depends on the existence of certain characteristics or building-blocks. These include supportive institutions, the capacity to innovate and the capacity to interpret and analyse knowledge in order to make decisions, and the ability to integrate uncertainty in planning and decision-making, for example. Important to note is that the five characteristics of adaptive capacity as set out in the LAC framework are not stand-alones from which one or more can be selected for attention. They shape and depend on each other.

Taking adaptive capacity on board does not mean adding five sets of each intervention for the five characteristics. It means understanding these dimensions of people's and communities' lives, and designing and implementing interventions in ways which enhance the way in which assets, institutions, innovation, knowledge flows and decision-making contribute to increased agency and more informed decision-making for the long term.

Tearfund has developed a framework bringing together activities focusing on adaptation to environmental degradation and climate change, DRR and development, highlighting the overlaps between sectors and projects (Wiggins and Wiggins 2009). Practical Action has developed a Vulnerability 2 Resilience framework,[4] while Marcus Oxley (Global Network for Disaster Reduction) has elaborated Consolidated Local Governance Rules and Principles (Oxley 2011). Different agencies use different language for what is the 'end goal'. For some, it is 'adaptive capacity', for others it is 'resilience' and for others still it is 'sustainable development'. Whatever the end goal is that is vocalised, there is growing convergence over what the key characteristics are that need to be strong in a community and beyond in order for any or all of the end goals to be achieved.

How and why development interventions support – or fail to support – adaptive capacity

The following sections are based to a large extent on research carried out under ACCRA in eight sites in Ethiopia, Uganda and Mozambique (see Tables 3.1, 3.2, 3.3) and highlight where rural development projects support – or fail to support – adaptive capacity. It is important to note that none of the interventions in the ACCRA research sites were designed specifically as 'adaptation interventions', but lessons can nevertheless be learned that are relevant for CBA, both in terms of what needs to be done to support the adaptive capacity of individuals, households and communities, but also what is required for scaling up such interventions.

Over-reliance on assets

Community-based projects studied as part of the ACCRA research, but also those implemented by other development partners, typically include a wide range of interventions including support to irrigation infrastructure, women's savings groups, agricultural extension, income diversification, the provision of assets (e.g. improved seeds, credits, social infrastructure) or training. Usually, these interventions focus on technology dissemination, addressing immediate needs raised in a 'participatory' process by the community, delivered through groups formed by the project and often including direct asset provision. They aim to increase income by providing, strengthening and diversifying livelihoods and assets, and capacity building of village-level organisations. Often, rural communities do not receive isolated projects: they receive support from government and non-government actors, including both 'discrete' projects and state services such as agricultural extension or safety nets.

The assumption behind providing assets – echoed by beneficiaries – is that assets can be converted into income and that this is making households more resilient to future shocks and stresses. This, however, rests on a set of further assumptions: (1) the people the project wanted to help have the labour, literacy, land, etc. to convert the assets provided into an income stream; (2) that groups which are often formed as part of development interventions are able to continue to work harmoniously in the future (e.g. no elite capture, no corruption, etc.) and carry out all functions as envisaged; (3) the technical information and inputs provided are the right ones for the specific land in that specific village and appropriate to the individual needs and constraints of each group member; and (4) that their appropriateness will remain despite changes in prices, demand, weather and climate, population pressure, land degradation, etc.

Like many community-based projects, few of the project interventions researched by ACCRA explicitly set out to support households and communities to respond to *future* development changes or climate pressures, but attempted to address immediate needs. If assets were considered as part of wider adaptive capacity, quite different project decisions might have been made, particularly also in view of scaling up such interventions beyond the immediate community where interventions are concentrated. For example, if we consider that the assets and technologies people use will need to change with changing circumstances, then projects, instead of handing out specific assets and technology, could put people in touch with sources of a range of technologies – and with sources of information which would help them to choose the most appropriate technology for their specific and changing circumstances. In other words, instead of handing out (for example) new seeds, projects could support linkages between farmers and seed suppliers and sources of new varieties (research stations, other farming communities, etc.), so that they can choose their own seeds in the future. Similar arguments can be made about projects handing out other kinds of assets, e.g. aimed at livelihoods diversification. There are examples where development partners aim at developing interventions that don't deliver assets but that put stakeholders with different knowledge systems in touch with each other, such as Tearfund's Climate Change and Environmental Degradation Risk and Adaptation Assessment (CEDRA) training, which was rolled out in Uganda through a cascading system involving staff of partner organisations at central and district levels aiming at enabling them to identify climate and environmental hazards from both scientific and community sources, how to prioritise hazards and how to develop appropriate adaptation options.

No assets without appropriate institutions

Assets alone are of not much use – they need a set of appropriate institutions to turn them into income streams. Irrigation infrastructure, for example, will only deliver water if there are institutions that ensure this. Women in one of ACCRA's research sites in Ethiopia (see Ludi *et al.* 2011), for example, complained that they did not receive their fair share of water because they lacked both money to pay the necessary bribes to those responsible for distributing water and also lacked the social status to claim their rights. So whatever projects do to support people to better adapt to climate – or other – change, if these interventions are not accompanied by the right institutions the chance that they might fail completely, or that they fail specific groups within a community, is considerable. Supporting institutions or even institutional change is a long-term process. Many of the projects studied as part of the ACCRA research were short-term – not by choice, but because donor support was, in some cases, provided for 12–18 months only. Only where development partners were able to engage over the longer term, more meaningful institutional change could be observed. An example is a project implemented in Niger where engagement with pastoral communities was possible over a period of more than 20 years, which enabled the building of sufficient trust between communities and project staff that ultimately led to changes in norms allowing women to run their own businesses and girls to attend school (Woodke 2007). Such interventions that challenge deep-rooted social and cultural norms are considered crucial to help prepare communities to cope with change, whether change relates to climatic factors, environmental degradation, demographic shifts or economic development. Besides the short time frame of many interventions that do not allow sufficient engagement with institutions, ACCRA research found three additional areas where lacking institutions were at the heart of unsustainable interventions: first, institutions can be subject to elite capture and corruption. Second, even where new institutions were established, they had not necessarily survived because they were introduced from outside and were not socially rooted; introducing new savings groups without considering existing institutions, rules and regulations and local social norms is an example. Third, technical interventions were introduced without considering the required institutional arrangements; natural resource management of common property is an example of where insufficient attention was paid to institutional arrangements and, as a result, these interventions often failed to achieve the stated goals. A key requirement for successful project interventions that address institutions and power relations is a long time horizon.

The crucial role of innovation for adaptive capacity

Being able to adapt to climate change depends a lot on how a culture supports innovation. Innovation is a key ingredient for adaptation to change for a number of reasons: first, in terms of climate change, we don't have all the answers yet. Models are predicting temperature increases but are not in agreement about how much and by when. Likewise, there is considerable uncertainty in terms of future rainfall: Will it be more? Or less? Falling at different times or in different amounts? Second, climate change, as stated before, is just one of the many changes people have to deal with. All this uncertainty means people constantly face new challenges. But these challenges are not the same everywhere. Thus, there is not one single solution or a specific best way to adapt to change – because its impacts are very different from place to place, but also because the vulnerability of people and communities or their capacity to adapt is different across different locations. Third, there is no end-point to which to adapt. What people need is the ability to adapt to change, now and in future, in a series of iterations. This means that people don't need hand-outs of new technologies or practices, but they need the capacity to innovate in sustainable ways.

ACCRA research in the project sites in Uganda, Ethiopia and Mozambique found that innovation was rarely considered by project interventions, whether they were intended to protect livelihoods, support or diversify livelihoods or manage disaster risks. In some villages, innovation was clearly constrained by a dominant culture which frowned on doing things differently. This culture is not challenged by development partners; on the contrary, it can be seen as being supported by introduction of an 'approved' innovation by external authorities. Opportunities are being missed to find out where, how and by who local innovation is happening, what the forces are that constrain people from innovating, and seeing how these could be addressed and innovation supported. Those implementing projects (and often researchers as well) all confused the provision of a new technology (an innovation) with supporting innovation as part of adaptive capacity. What is required is a shift in thinking of development partners – away from providing technology or innovation towards supporting an innovation process which requires a combination of technologies, information and knowledge and new forms of social organisations and institutions. Being innovative can be constrained by a number of factors, among them cultural barriers, the ability to bear financial risks, lack of information and knowledge or lack of confidence. All these could be addressed by development partners if they were thinking of development as supporting adaptive capacity and not as mainly providing assets.

Successful adaptation – which will imply for many transforming their current livelihoods – and successful scaling up of interventions will mean that a lot of innovation is required. This also means that projects need to completely change their way of working, moving away from an understanding of static rural contexts and from planned small-scale activities, concrete and tangible deliverables and milestones to a much more 'fuzzy' approach, where neither the end-point, the concrete deliverable nor the specific approach to get there is exactly known at the outset of the project. This will also mean that many development interventions will have to move away from a community-based, often sectoral focus with tangible deliverables to a systems focus that considers interlinked scales, sectors and dimensions spanning from assets to forward-looking planning and governance systems such as those described in the LAC framework.

Translating information into knowledge

To be able to adapt to change or to innovate requires having access to information. This is not only relevant for people or communities, but for projects and governments alike. ACCRA's research in Ethiopia, Uganda and Mozambique found that governments' and projects' treatment of information was largely confined to providing technical packages. Also, the assumption is that whatever technical information is provided is 'correct' and appropriate – for example, technologies maximising yield or income for a standard set of circumstances. However, seeing information, and the process of transforming information into knowledge, as key components of adaptive capacity and scaling up interventions would encourage us to put far more emphasis on giving people a much wider range of information appropriate to a much wider range of circumstances. Information is needed to deal with the process of change as well as to inform decisions about adapting to future circumstances. This information should cover both shorter- and longer-term futures, and needs to recognise the associated uncertainties involved. Crucially, this means turning information into knowledge – supporting people's ability to use the information for decision-making, particularly difficult when predictions are uncertain. A crucial pre-requisite for this is to establish networks and bring the diverse stakeholders with their specific knowledge sets – farmers, researchers, academics, government officials, NGOs – together to identify sustainable and effective responses.

Supporting adaptive capacity means flexible and forward-looking planning and decision-making

There will be change – climate and otherwise – but in which direction it is going, how big it will be, how much it will affect current livelihoods and practices and whose livelihoods, is all uncertain. This uncertainty may lead to decisions being taken that increase the risk of maladaptation, i.e. decision-making that leads to long-term increases in vulnerability. Two sources for this have been identified: first, climate information is misinterpreted and the uncertainty related to climate change carries the potential for ill-informed planning. Second, many interventions are being made without considering any evidence about longer-term climatic changes or without an analysis of any other change, whether economic, political, demographic or otherwise. A prominent example is the promotion of small-scale irrigation in many semi-arid areas as a climate change adaptation intervention, but also as a means of diversifying households income sources and ultimately of increasing incomes. At the current moment and with current knowledge, it is impossible to know whether these interventions represent adaptation to climate change or maladaptation. There is a lack of information about the future scale of irrigation and about the long-term availability of water, the feasibility of irrigation from an economic and institutional perspective, and the wider environmental, social and economic impacts.

ACCRA's research in the eight project sites in Ethiopia, Uganda and Mozambique did not find that this information was being sought or these wider questions discussed, but rather found that assumptions were made that might be relevant for the short term but that might be very wrong in the longer term. This does not only highlight the need for having and using information, but also for decision-making to be forward-looking. Given that we know the future is uncertain, both in terms of future climates and development, rigid, centralised planning runs high risks of either failing to respond to changing circumstances or even leading to maladaptation. ACCRA's research suggests that there were two problems to decision-making and planning: first, planning is not incorporating either available knowledge of climate or considering other developments such as world food prices, demographic changes, etc., but is primarily reactive and focused on addressing immediate local needs. In part this stems from current modalities of 'participation' (asking 'communities' what they want), from short-term funding modalities and from a projectised approach to development working at the small scale and from plans, but in the absence of coherent long-term strategies. Second, there is still a reliance on top-down planning which does not support local flexible decision-making and agency. Both of these problems are centrally relevant for successful scaling-up of interventions that would allow people to adapt to change, which requires that pervasive drivers of vulnerability are addressed in the planning process – and this requires more often than not a focus beyond the local level.

Conclusion: enhancing the adaptation–development relationship through integration

Research carried out under ACCRA showed that development interventions typically focus on technology dissemination, addressing immediate needs raised in a 'participatory' process by a 'community', delivered through groups formed by the project and often including direct asset provision. Such interventions, whether initiated by Sustainable Livelihoods, Social Protection or DRR programmes did contribute to strengthen or build assets, and in some situations institutions, but they failed to adopt a broader view of what is required to support local people and communities to build their adaptive capacity to change. One missing piece was found that development partners did not adequately take into consideration how to support people's *agency*. Without agency there is no adaptive capacity. Development interventions insufficiently reflected on and analysed their interventions in view of whether or not they contributed to support the ability of individuals, households or communities to respond to changing circumstances, whether weather-related or otherwise. Development interventions focused on providing outside assistance rather than enabling people to access assets, information, skills and knowledge that would allow them to shape their own future.

Development means supporting people for a future with change – uncertain change. It is not possible to plan centrally for climate change with any certainty, partly because no one knows the exact way in which climate will change locally and partly because all the impacts of climate change on people's lives are influenced and felt through a host of other changes (changes in access to natural resources, prices, population movements, etc.), mostly with drivers beyond the local scale. Irrigation channels alone do not help prepare people for 'climate change'. Enhanced adaptive capacity does. This might be in part because of a results-driven agenda and a more recent approach on 'value for money', which makes it easier for development partners to deliver assets and training that can easily be counted rather than the more 'software' support required to support people to learn, experiment, share and shape their own environment and future.

Improving 'governance' is a core condition for support to adaptive capacity. 'Governance' is about the exercise of power in decision-making: it does not only refer to governments, but also to NGOs and other development partners, local authorities, local institutions, private sector actors and communities themselves.

An analysis of development interventions in the ACCRA case study areas showed that interventions were planned and delivered without taking into account all dimensions of adaptive capacity. In part this may be explained by a lack of focus on agency, which would recognise that it is not crucial under what sector or approach interventions are delivered, but how. A significant and profound change in thinking is required to improve the contribution of development interventions to supporting adaptive capacity at scale. This will make changes essential in everything from the skills needed by programme designers and implementers to changes in relations between local authorities and central ministries, or between implementing agencies and donors.

Although supporting adaptive capacity of individuals, households and communities is crucially important, it is not sufficient. Many of the factors that make people vulnerable lie outside their sphere of influence. Interventions that are aimed at supporting adaptive capacity in its broad sense thus need to tackle these drivers of vulnerability. That requires a paradigm shift of development interventions away for small, short-term community-based projects focusing on tangible deliverables to interventions that are able to support people and systems to take account of and address the multiple and interacting changes beyond climate change, that are able to deliver adaptive capacity at the local scale while also addressing the root-causes of vulnerability, and that are able to evolve in a constantly changing environment.

Many NGOs, community-based organisations (CBOs), donor agencies and government agencies are aiming at supporting interventions that contribute to building adaptive capacity of people, communities and systems, and there are positive examples where development and adaptation are being integrated. Experience from research and practice suggests that for integrating climate change adaptation into development practice with a view of supporting adaptive capacity at scale, the following lessons learned about how development interventions can best contribute to adaptive capacity are crucial:

- Project interventions need to be based on a thorough analysis of the longer-term future and of uncertain changes – including both climatic and other parameters.
- Development interventions need to assess their impacts on people's agency and how they support their adaptive capacity.
- A new approach to development is required that is not limited to handing out specific assets and technology, but would put people at the centre and put them in touch with sources of a range of information which would help them to choose the most appropriate technology for their specific and changing circumstances.

- Moving away from handing out assets and technologies implies a renewed focus on the importance of institutions. They define to what use people can put the resources to which they have access, how people interact and how collectively they make decisions about their future. Care is required, though, to understand institutions much better, build on existing rules and regulations and be aware that institutions can be subject to elite capture and corruption.
- A focus on adaptive capacity and innovation requires a shift in thinking – away from providing technology or innovation towards supporting an innovation process that requires a combination of technologies, information and knowledge and new forms of social organisations and institutions.
- Enabling people to innovate requires supporting people to access and use information and translate this into knowledge for decision-making that takes account of an uncertain future.
- Governance requires a shift away from rigid, centralised, short-term and small-scale planning based on local needs that runs the risk of maladaptation, and towards an evidence-based planning process informed by different scenarios of how a variety of climatic and non-climatic parameters might change over the next decades.
- Action is required by multiple agents at multiple scales – international, national, district and community level. This chapter has laid out key characteristics for supporting adaptive capacity in the form of the LAC framework. If taken seriously, the five characteristics discussed allow assessing – and designing – development interventions that allow for an approach integrating development and adaptation at scale.

Notes

1 When making reference to ACCRA, this chapter draws on: Ludi *et al.* (2012); Levine *et al.* (2011); and Jones *et al.* (2010).

2 ACCRA, the Africa Climate Change Resilience Alliance, is a consortium made up of Oxfam GB, the Overseas Development Institute (ODI), Save the Children, Care International and World Vision International, and is funded by the UK's DFID. The aim of the first phase of ACCRA (2009–2011) was to understand how social protection, livelihoods and DRR projects by ACCRA consortium partners build adaptive capacity in beneficiaries and how these approaches can be strengthened. It also aimed at increasing governments' and development actors' use of evidence in designing and implementing both humanitarian and development interventions that increase poor and vulnerable communities' capacities to adapt to climate hazards, variability and change, support local and national government to build capacity to implement interventions which can build communities' adaptive capacities and to encourage learning across countries and disciplines. The programme has studied projects (and to a lesser extent government interventions) initiated by the consortium partners in the fields of DRR, sustainable livelihoods and social protection in project sites in Ethiopia, Uganda and Mozambique to better understand how programmes and policies can support communities' and individuals' capacities to adapt to climate change.

3 Tearfund is a UK-based international relief and development agency and was one of the first development NGOs to recognise and respond to the adverse impacts of climate

change and environmental degradation on its development projects. This was in response to observations and requests for assistance from partner organisations in developing countries (i.e. local development NGOs) (Roach 2005). Tearfund has developed climate, environmental degradation and DRR resources and training (e.g. CEDRA (see Wiggins 2009) or PADR (see Hansford 2011)) to enable its partners to modify all development projects, as necessary, with the ultimate goal that all projects incorporate adaptation to climate change and environmental degradation, DRR and sustainable development. It is at the early stages of this goal.

4 Available at http://practicalaction.org/media/view/9654.

References

Agrawala, S., 2005. *Bridge over trouble waters: Linking climate change and development*. Paris, OECD.

Arnall, A., 2012 (forthcoming). *Preparing for the future in Mozambique: Understanding the influence of development interventions on adaptive capacity at the local level.* Africa Climate Change Resilience Alliance (ACCRA) Mozambique Synthesis Report.

Brooks, N., Anderson, S., Ayers, J., Burton, I. and Tellam, I., 2011. *Tracking adaptation and measuring development.* Working Paper. London, IIED.

Dodman, D. and Mitlin, D., 2010. Challenges for community based adaptation: discovering the potential for transformation. Draft Paper.

Hansford, B., 2011. *ROOTS 9: Reducing risk of disaster in our communities*. Teddington, Tearfund UK.

Jones, L., Ludi, E. and Levine, S., 2010. *Towards a characterisation of adaptive capacity: A framework for analysing adaptive capacity at the local level.* London, Overseas Development Institute.

Jones, L., Ayorekire, F., Barihaihi, M., Kagoro, A. and Ruta, D., 2011. *Preparing for the future in Uganda: Understanding the influence of development interventions on adaptive capacity at the local level.* Africa Climate Change Resilience Alliance (ACCRA) Uganda Synthesis Report. Available at http://tinyurl.com/ACCRA-Publications (accessed 14 February 2012).

Levine, S., Ludi, E. and Jones, L., 2011. *Rethinking support for adaptive capacity to climate change: The role of development – findings from Mozambique, Uganda and Ethiopia. A Report for the Africa Climate Change Resilience Alliance*. London, ODI. Available at: http://tinyurl.com/ACCRA-Publications (accessed 14 February 2012).

Ludi, E., Million, G., Wilson, K., Kindie, T., Beneberu, S., Levine, S. and Jones, L., 2011. *Preparing for the future? Understanding the influence of development interventions on adaptive capacity at local level in Ethiopia.* Africa Climate Change Resilience Alliance (ACCRA) Ethiopia Synthesis Report. Available at: http://tinyurl.com/ACCRA-Publications (accessed 14 February 2012).

Ludi, E., Levine, S. and Jones, L., 2012. *Changing focus: How to start taking adaptive capacity seriously.* London, ODI. Available at http://tinyurl.com/ODI-Briefing-Paper (accessed 14 February 2012).

Oxley, M., 2011. *A systems approach to building resilience in hazard-prone areas in a changing climate*. Global Network for Disaster Reduction.

Parry, M., 2009. Climate change is a development issue, and only sustainable development can confront the challenge. *Climate & Development* 1 (2009), 5–9.

Roach, R., 2005. *Dried up, drowned out: Voices from a developing world on a changing climate*. Teddington, Tearfund, UK.

Smith, J.B., Klein, R. and Huq, S., 2003. *Climate change, adaptive capacity and development*. London, Imperial College Press.

Wiggins, M. and Wiggins, S., 2009. *Environmental assessment for use on medium-impact projects*. Teddington, Tearfund UK.

Wiggins, S., 2009. *CEDRA: Climate change and environmental degradation risk and adaptation assessment*. Teddington, Tearfund UK.

Woodke, J., 2007. The impact of climate change on nomadic people. *Footsteps* 70. Available at: http://tilz.tearfund.org/Publications/Footsteps+61-70/Footsteps+70 (accessed 19 September 2013).

World Bank, 2010. *World development report: Development and climate change*. Washington, DC: World Bank.

Part II

Making it happen

Finance, frameworks and tools for scaling up CBA

4 The economics of CBA

Muyeye Chambwera, Maria Rebecca Campos, Martin Obermaier, Jessica Frank and Denise Loga

Adapting to climate change presents itself as an economic problem because it addresses the bigger problem of allocating scarce resources to attain sustainable development (Chambwera and Stage 2010; Stern 2006). At the community level, adaptation involves different combinations of collective and individual actions. These depend on private household resources or public resources, or both at the same time. The decision to allocate resources for adaptation is based on formal and informal considerations of the value or benefits of the actions undertaken against their costs, i.e. adaptation decisions may be viewed as investments that have direct or indirect benefits now or in the future. For example, when farmers invest in soil and water conservation measures, they anticipate an increase or stabilisation of crop yields in the face of adverse climatic events such as droughts or floods. Similarly, farmers who invest in floating gardens expect to maintain a stable flow of vegetables through all seasons. The extra effort they use, or the additional expenditure they incur, affects their existing livelihood pursuits. For example, the extra labour put into soil and water conservation or constructing floating gardens could have been used to earn an income somewhere or for leisure. The same applies to community adaptation activities such as planting trees in degraded catchments to reduce erosion and siltation of rivers or dams.

It is reasonable to assume that, even though firms and households will consider the need to adapt to climate change in their overall strategies, they will not do this in isolation from other decisions. Households, like firms, will presumably seek to maximise their profits regardless of the weather, and climate change will be only one of the many factors affecting their production and investment decisions. This makes the economic considerations highly complex, in that adaptation actions are not considered in isolation, but as part of broader livelihood strategies.

The informal nature of economic considerations made by households and communities, together with the broad nature of the decision-making framework (climate adaptation is part of a larger welfare function rather than a stand-alone consideration), makes the practical dimension of the economics of CBA complex and often difficult to subject to formal economic criteria such as straightforward cost–benefit analysis. This is complicated by the fact that households and communities do not always buy adaptation goods and services on the market, and these goods and services are not always expressed in monetary terms. Rather, communities adapt using combinations of purchased and non-purchased goods and services, cash and in-kind transactions, and sometimes use inputs that are often undervalued, such as labour. Equally, the consideration of adaptation is integrated into the decision-making processes of households and communities incrementally as part of decisions to cope with climate variability and extreme events.

These considerations suggest the need to use both formal and informal economic approaches in community-based adaptation (CBA). Formal economic considerations as used here refer to the use of economic functions such as cost–benefit, cost-effective, multi-criteria functions that use decision rules such as benefit–cost ratios, net present values and profit–loss estimates using quantitative and qualitative data to evaluate adaptation options. Informal economic considerations are integrated in the livelihood choices that households and communities make, in which managing climate change impacts is one, and they are often informed by experience, availability of resources as well as other social, institutional and political factors. Catchment management, for example, could be influenced by village boundaries, the need for communal grazing and other considerations.

Scaling up adaptation actions involves the wider uptake of actions that have been tried and tested in limited locations. Where external resources are available – for example, from donors or government – formal economic approaches may be used to evaluate specific options. Where external resources are unavailable or insufficient, the formal and informal economic considerations made by households individually and collectively are key determinants to the scaling up of pilots, among other non-economic considerations. For households and communities, the capacity to undertake autonomous adaptation will be constrained by a number of factors outside their immediate control, such as publicly provided infrastructure, information, access to local and external markets, technology and access to credit, among others. When these are in place, it is reasonable to assume that households and communities will implement adaptation actions that have net positive welfare effects (in monetary and non-monetary terms), leading to their scaling up.

Climate change adaptation is also enhanced by other approaches, processes and players that interact with communities. These include the private sector, market involvement, risk management schemes such as insurance and natural resource-based incentives that contribute to adaptation. This chapter presents the role of specific economic tools and the enhancing approaches based on community-based case studies to highlight their applications. The chapter concludes by discussing the limits of the use of economics in CBA.

The use of economic tools in adaptation decision-making

Decision-makers such as communities, donors, governments, NGOs and investors need to base their decisions on several criteria, including economic criteria, leading them to employ tools that assess the investment-worthiness of adaptation actions. They also need to assess whether pilot actions could be implemented at larger scales using economic information among others. The most common tools that are applied for these purposes include cost–benefit analysis, cost-effectiveness, and multi-criteria analysis.

Cost–benefit analysis

Managers and analysts in both the government and private sectors are often asked to provide analysis in support of resource allocation decisions confronting them. Difficult choices are involved when resources are scarce. Thus, financing new programmes can be favoured at the expense of existing ones. This is also true in the case of management interventions to CBA regarding climate change. Regardless of who makes the decision, what varies is the level of investment and the resources at stake. This tool, used by decision-makers, is a systematic process of computing and comparing the benefits and costs of a project. It determines whether the benefits outweigh the costs of a project and by how much, adjusted for the time value of money. After identifying the benefits and costs of the project, financial analysis is conducted to arrive at the following indicators: net present value or worth (NPV or NPW), internal rate of return (IRR), payback period and benefit–cost ratio (BCR). The stream of benefits and costs is computed for the duration of the project using a discount rate in order to get the NPV. The choice of discount rate is rather subjective, but could make a large difference when assessing interventions with long-term effects like those affecting climate change. A small discount rate values the current generations equally to the future generation. Large discount rates mean that money today is valued more than it will be in the future. IRR is the rate of interest that equates the present value of costs with the present value of benefits. The payback or pay-off period is the length of time (e.g. years) it will take for the project to pay for itself or when the stream of benefits is equal to the stream of costs. The BCR is computed by dividing the sum of discounted benefits by the sum of the discounted costs. The decision rule is that the project is feasible when:

- BCR > 1
- NPV > 0
- IRR > bank interest rate
- payback period should be within the project life.

In some cases, either all or only some of the criteria are met so the projects must be ranked and prioritised.

The impact of intangible benefits and costs cannot be assigned a monetary value in financial analysis. These intangibles must be clearly identified in the cost–benefit analysis. Economic analysis values these intangibles. The decision criteria in economic analysis are the same with financial analysis. Sensitivity or risk analysis needs to be done where uncertainty in the cost (for example:

increase in wages) or benefits (for example: increase in selling price) items are assumed.

Cost-effectiveness analysis

When it is difficult to put monetary values to benefits, cost-effectiveness analysis (CEA), also known as cost-minimisation technique, is utilised. For example, because of more frequent floods as a result of climate change, there might be two highway upgrade options that will result in the saving of lives. The decision is based on choosing between the options on the basis of minimum cost. CEA is used in coming up with a decision in the planning and management of many types of organised activities. Thus, it is applicable to CBA to climate change.

CEA is used mostly in the field of health services, where it may be inappropriate to monetise the health effect. In CBA to climate change, CEA can be used to compare different medications to combat flood-induced diseases as a result of climate change in a community. CEA compares the relative costs and outcomes (or effects) of at least two options. Here, the effects can be measured by the number of people who have been cured of the disease and the number of symptom-free days experienced by a patient in both options. The economic efficiency indicator for CEA is expressed as a ratio where the numerator is the cost associated with the health gain, while the denominator is effect or the gain in health from a measure (years of life, premature births averted, sights regained).

Multi-criteria decision analysis

Managers and decision-makers often encounter structuring and planning problems involving multiple criteria where a unique optimal solution does not exist for such problems. Thus, it necessitates using the decision-makers' preferences to differentiate among options or solutions. Multi-criteria decision analysis (MCDA) is applicable in information economic analysis where weights are assigned to both tangible and intangible economic benefits and costs/risks. It can also be used as a decision tool for projects where both the economic and non-economic benefits and costs/risks are used for ranking and prioritising.

MCDA in information economic analysis provides a common ground for evaluating all the factors in coming up with a decision by assigning weights to both the benefits and risks attributed to the solution. All the options are weighted and evaluated. The option with the highest weight is the recommended option.

MCDA is conducted by forming a committee to evaluate the solution. This committee comprises managers affected, directly or indirectly, by the new system. The committee determines the intangible costs/risks and benefits of the new system. This may be based on a list of suggested intangible costs and benefits, provided by the analyst preparing the cost–benefit analysis. The decision is based on the importance of each risk and benefit. A weighting factor is assigned to each intangible risk, tangible benefit and intangible benefit. This weight is represented as a percentage, where the sum of all of the weights is 100 per cent.

The estimate of the probability of realising the benefit or incurring the risk translates to a ranking on a scale of zero to five, where zero is no probability and five is very high probability. The risk is assigned a negative ranking, where zero is no risk and minus five is a very high risk. The option is evaluated by multiplying the weight of each benefit or risk by its probability ranking. These numbers are added to obtain the final score.

The same procedure is followed for projects or management options involving multidisciplinary decision-making.

Case study: application of cost–benefit analysis in the Philippines

Climate change has affected the sea cucumber industry of the Philippines, which is the world's second major producer and exporter of sea cucumbers. Moreover, this has reduced income for the country's economy, the commercial fishing sector as well as marginal fishermen who rely on it as their source of livelihood. Three types of sea cucumber fishing techniques are: commercial fishing; harvesting sea cucumbers as by-catch; and gleaning. This research focused on gleaning because the fishermen have no control of the shallow coral reef flats where they catch sea cucumbers. Collection is done by small-scale or artisanal fishers, involving men, women and children. This activity is carried out during low tide in shallow intertidal reef flats. Gleaning is often classified as 'informal' work acting as a safety net for the rural landless. The community of sea cucumber fishers in Sorsogon organised themselves and came up with three coastal management options. Cost–benefit analysis was applied to the three options to assess the best among them. Financial cost–benefit analysis (considering only the direct costs and benefits of the project from the standpoint of an investor) and economic cost–benefit analysis (which takes into account non-market values such as existence values and services such as cultural values (existence, research, etc.) and regulating services (coastal protection, waste assimilation, etc.)) were undertaken.

Option 1: stand-alone MPA

Ideally, a marine refuge and sanctuary (popularly known as an MPA or Marine Protected Area) is about 15–30 hectares, kept as a no-take zone, with markers delineating the area. A 50–100 m buffer zone in most cases is added around the MPA where controlled subsistence fishing is allowed for members of the MPA community-based team. Considered as the most effective community-based method to keep coral reefs and sea grass beds healthy, it has even worked well for the restoration of damaged reefs. The establishment of MPAs in the Philippines is justified by the Philippine Fisheries Code, which stipulates that 15 per cent of municipal waters must be declared as a marine refuge and sanctuary.

Option 2: MPA–artificial reef (MPA–AR) model

In some areas, artificial reefs are introduced into the MPA to enhance the recruitment in corals and provide shelter for reproducing fish. This is commonly introduced in highly degraded reefs. Apart from not being natural, a primary disadvantage of artificial reefs is the high cost of investment. Further, since the composition of the artificial substrate may not be as homogeneous as the composition of the natural substrates, the recruitment rates of coral larvae may be lower. On the other hand, they have served as sources of attraction for tourists, which may generate income that could offset the cost of installation and maintenance.

Option 3: MPA–sea ranch (MPA–SR) model

A 2–3 hectare sea ranch is established next to a 16 (or so) hectare MPA to provide ecotourism services, which is the most commonly used financial model among the successful MPAs. Under this model, the MPA and sea ranch benefit each other in at least three ways: protection from poachers; ecological stability; and climate protection, as well as conservation. But the establishment of the basic services needed for ecotourism, such as hotels, restaurants, roads and accessibility are costly, while newly built MPAs would require some time for the reefs to recover to a state that tourists would enjoy viewing. This early stage of protection spans around 3–5 years. Within this very critical period the ecotourism model could not be used to provide financial support to the MPA's management. Thus economic analysis is necessary to establish viability before upscaling.

The results of financial and economic analyses of the three options are presented below. The estimates were based on a 20-year assumed lifespan of the project. The discount rate is 18 per cent, the prevailing interest rate of government-run banks such as the Landbank of the Philippines.

The *stand-alone MPA* and *MPA–AR* options are not financially feasible, while the *MPA–sea ranch model*, on the other hand, is highly profitable as a management option (Table 4.2).

The *stand-alone MPA* and *MPA–AR* management options are still not economically efficient, even when benefits for waste assimilation, coastal protection and existence values, etc., are factored in. These types of projects require subsidies from government funds, or grants from aid agencies or the private sector.

Table 4.1 Financial efficiency indicators of three types of MPA

Type of MPA	*NPV (Philippine peso or PhP)*	*IRR (%)*	*Payback period (years)*	*BCR*
Stand-alone MPA	–2.5 million	Non-existent	21.06	<1
MPA–AR	–3.8 million	Non-existent	56.69	<1
MPA–SR	669,954.20	39.9	1.16	1.724

Table 4.2 Economic efficiency indicators of three types of MPA

Type of MPA	*NPV (PhP)*	*EIRR*	*Payback period (years)*	*BCR*
Stand-alone MPA	–2.4 million	N/A	9.28	0.068
MPA–AR	–3.7 million	N/A	31.7	0.058
MPA–SR	764,384.19	44%	1.08	1.68

This is perhaps the primary reason why only 33 per cent of all the MPAs in the Visayas have been successful (Alcala *et al.* 2008). The *MPA–SR* model becomes even more attractive when non-financial values are incorporated in the analysis. Using the results of this study as a guide, the *MPA–SR* model is now being implemented in Sorsogon, the project site.

As shown in the Philippines case study, economic analyses must consider both private and public values of costs and benefits, i.e. financial and economic analyses. Several adaptation strategies must be developed and analysed before scaling up any one of them, and sustainable revenue mechanisms must be put in place to offset costs that cannot be avoided. In addition to cost–benefit analysis, other economic tools can also be employed depending on the type of decision that needs to be made by specific stakeholders.

The role of markets and farmer organisations in scaling up community adaptation

Markets provide a means for goods and services to be exchanged by buyers and sellers. Households and communities are also fully or partially engaged in local and external markets (national, regional and global). Market engagement can be direct or indirect, sometimes participating in external markets through intermediaries and agents, such as in the export of specialised commodities like flowers, coffee, or in the marketing of community tourism products.

Although there are no markets primarily designed to provide adaptation goods and services, markets for several existing goods and services can play a role in building local adaptive capacity as local communities can use their respective comparative advantages and specialisations to cushion themselves from the impacts of climate change. Locally, livestock farmers whose pastures are depleted by drought may trade their livestock for grazing rights with farmers who have fallow land in wetland areas. Access to markets also enables farmers to grow cash crops that earn them an income that could be invested in assets that build adaptive capacity, such as boreholes or solar equipment. Diversification away from food crops to cash crops broadens their livelihood base.

When markets are functioning well and reach to vulnerable communities they can be effective means for distributing adaptation goods on a large scale. For example, new crop varieties, information technologies like mobile phones and irrigation equipment can reach local communities at scale through market channels. Governments can also use market channels to subsidise adaptation goods and services without being directly involved in the distribution.

The potential for markets to scale-up adaptation depends on their ability to serve the needs of the very poor and vulnerable communities, with pro-poor markets having greater potential to support the upscaling of CBA. The Adapta Sertão project in Brazil demonstrates how the development of pro-poor markets contributes to adaptation in local communities.

Case study: adaptation through pro-poor market development in Brazil

Adapta Sertão[1] is a CBA project aiming at poor family farmers in the semi-arid interior of northeast Brazil, located around the municipality of Pintadas, some 300 km northeast of Salvador. Family farmers in Pintadas rely mainly on small-scale livestock raising, milk production and crops such as cassava, beans or corn for subsistence, but ongoing degradation of the native Caatinga vegetation due to extensive cattle raising and deforestation has long since undermined agricultural activities, a particularly serious problem as many farmers continue to lack access to productive land. These vulnerabilities are exacerbated by insufficient market development, including widespread inadequate production techniques, limited technical assistance and difficult market access. Even where farmers have been willing to invest in technology, private firms often hold a monopoly on sales of irrigation and other production technologies and operate in distant urban centres. This has simultaneously affected local availability of appropriate technologies and their economic viability, given the considerable mark-up on system costs. In this context, climate change constitutes an additional layer to existing socio-ecological vulnerabilities by potentially reducing already scarce rainfalls and increasing the frequency and length of drought periods.

Dissemination of appropriate technologies and agricultural practices through market-based instruments is a central component of Adapta Sertão's approach to strengthening the adaptive capacity of farmers. In a participatory process, local farmer associations or cooperatives are capacitated to become retailers and technical assistance providers of production technologies and inputs such as irrigation systems, water pumps or quality seeds. The systems are designed and implemented by the cooperatives' technicians and get to the farmers at a considerably lower price as contracts with the technology manufacturers guarantee cooperatives to become local distributors, thus eliminating any middlemen. Farmers themselves receive funding to implement their systems from available local and federal micro-credit schemes. The advantage of relying on local cooperative or farmer association members for this is that they often have the trust of their farmer associates. For a majority of farmers drip irrigation has been shown to be advantageous: it reduces production losses due to drought, creates higher value-added produce (often increasing family income) than under rain-fed conditions and allows for a more balanced family diet. The local cooperative in Pintadas also helps to access the market by setting up a market stall with 40 associates at the city's weekly market. The cooperative has also managed to sell the farmers' products through Brazil's recent Food Acquisition Programs (PAA and PNAE), where the federal government buys small farmer products from cooperatives or associations at fixed prices without competitive bidding processes for consumption in local or regional public institutions such as hospitals or schools.

By providing an impetus for the development of local markets, Adapta Sertão's approach to capacity building seems to benefit both farmers and cooperatives alike, and may thus make the project investments more sustainable, even in cases where project funds run out. In this, the scaling up of Adapta Sertão's current project experiences continues to understand adaptation as a process which necessarily needs to create adaptive capacity at different scales. The recent adaptation forums organised in Pintadas thus invite policy-makers, representatives from development agencies and NGOs, and cooperative members or farmers. These forums provide a venue where barriers for sustainability and scaling up are discussed.

Communities access markets as individual households or as groups. Organised community groups in developing countries are most common in agriculture, where farmer organisations enable small farmers to access specialised and distant markets at scale, thereby reducing transaction costs associated with accessing such markets. Organised groups are also a cost-effective platform for disseminating information and building the capacity of local communities, which can be applied to adaptation. Fairtrade provides a good example of the role of community access to markets through farmer organisation at scale. The growth in both sales of Fairtrade products, mostly produced by smallholders in developing countries, as well as the growth in the numbers of smallholder farmers involved demonstrates the role of organised market access in scaling up market-based agricultural enterprises. Between 2009 and 2010, the retail sales value of Fairtrade-marked products rose 40 per cent from £18.6 million to £1.17 billion. Small farmer organisations capture two-thirds of the UK Fairtrade market while individuals from least income countries make up 39 per cent of the 915,244 individuals from Fairtrade-certified producers supplying the UK market (Fairtrade Foundation 2011).

As well as providing economic benefits such as premium prices, farmer organisations are also channels or institutions through which to deliver CBA to local communities. They are useful platforms for raising awareness, influencing, learning and generating extra resources for adaptation. Owing to these market links, farmer organisations are able to tap into value chain funding for adaptation as well as traditional grant funding, pricing premiums plus non-monetary resources of different stakeholders within their supply chain. As businesses, farmer organisations also have the capacity to handle and distribute significant resources to their large memberships.

There are limits and possible negative effects of markets on community adaptation and scaling it up. Involvement in external markets could lead to excessive specialisation, narrowing down the livelihood base of communities such that they are more exposed to climatic impacts. Access to external markets could also lead to very poor and vulnerable groups being priced out of resources such as land by wealthier members of society, leaving disadvantaged groups even more vulnerable. Markets could also lead to commoditisation and de facto privatisation of communal resources, effectively reducing the numbers of community members benefiting from them. For this reason, community adaptation should not be left solely to market forces as unregulated market activities could be a form of maladaptation.

Insurance

Climate variability and change exposes poor and vulnerable farmers, as well as urban dwellers in developing countries, to unpredictable risks of disease, food insecurity and damages to assets and livelihoods, which creates the need for corresponding risk management instruments (Intellecap 2010).

Innovative insurance models are at the forefront of private sector engagement in climate change adaptation at the community level, where respective safety nets and protective mechanisms are significantly lacking. Donors and development practitioners are increasingly changing by investing more resources in financial instruments pre-catastrophe and fewer in post-catastrophe assistance (Suarez and Linnerooth-Bayer 2011).

Index-based weather insurances are currently applied risk management tools for smallholder farmers to protect their livelihoods in case of yield and other losses as a result of extreme events and climate variability and change. Through defining an index, farmers are insured against the case that the required weather conditions (e.g. a certain amount of rainfall) do not occur or that an extreme event (e.g. drought, flood) leads to a significant reduction in agricultural production or climate-related loss and damages (Suarez and Linnerooth-Bayer 2011).

Successes

In order to provide a protective mechanism against climate-induced hazards to agricultural productivity, like changing rain patterns in terms of time and amounts, prolonged dry periods and extreme events, micro-insurance products provide a potential adaptation solution by alleviating local vulnerability (Suarez and Linnerooth-Bayer 2010).

Challenges for scaling up

In terms of financial profitability, insurance providers are often not able to generate a service that fulfils the needs of the most poor and vulnerable people with almost non-existent purchasing power, who are excluded from the existing market for insurance products. Therefore, support from donors or international finance institutions is almost always required (Intellecap 2010). These subsidies not only limit sustainability but may also lead to maladaptation by encouraging smallholders to continue with agricultural production while not taking into account more viable income diversification opportunities (Suarez and Linnerooth-Bayer 2010).

So far many diverse insurance models for adaptation exist; however, they are only implemented in the form of pilot projects, and the opportunities for scaling up are limited. In order to achieve more comprehensive developmental outcomes they would require mainstreaming in national and global development agendas, which so far has not been achieved (Suarez and Linnerooth-Bayer 2010).

Case study: drought insurance in Malawi

In Malawi most of the poor people are excluded from financial and credit services, which significantly limits their opportunities for securing and expanding their asset base and engaging in adaptation and climate-related risk management, such as by buying more climate-resilient seed varieties.

The pilot case in Malawi offered a special drought insurance product to farmers that was combined with a loan element. Additionally, it was not framed around covering a proportional loss but would protect against the changing weather patterns and climate events that would lead to the economic and social damages – for instance, inadequate precipitation throughout the cultivation cycle. The weather insurance consisted first of all of a premium calculated based on the likelihood of a payout, and second on a credit component that enabled smallholders to purchase better inputs. Moreover, individual prices were operated based on seasonal precipitation projections including the El Nino Southern Oscillation (ENSO). Farmers had to pay once per season an amount equivalent to their input expenditures additional to the premium, interests and taxes, which they could afford due to the higher revenues generated through an increase in yields. Only smallholders who were members of the National Smallholder Farmers Association of Malawi (NASFAM) were able to participate in the scheme, which was entirely administered by the cooperative and operated under group dynamics, such as security provision and loan repayment obligations.

The case study has revealed several beneficial outcomes that led to increased adaptive capacity at the community level. Smallholder farmers were able to increase their output by twice as much during preferential weather periods, which also enabled them to save some of their earnings for seasons with worse conditions. Thus, the insurance contributed to the reduction of vulnerability related to food insecurity (through increased outputs during preferential cultivating conditions) while at the same time alleviating the actual yield loss (through improved agricultural practices). Furthermore, the management expenses of this insurance product related to payouts were less than those of usual crop insurances due to the fact that payments are merely connected to the surveillance of weather data instead of occurred damages. Since payments are also not related to smallholders' individual farming strategies and the fact that the weather data information is equally accessible for the service provider and the beneficiary, the risk of moral hazard is significantly reduced. The Malawian case also presented a case for achieving financial sustainability, because as soon as the scheme was set up, the improvement of yields covered the remaining ongoing expenditures for the insurance premiums (Suarez and Linnerooth-Bayer 2010).

Drawbacks of weather insurance in general are that payments may not adequately correspond to the yield losses. Moreover, basis risk is another challenge to face as farmers who experienced the climate condition that coincided with the data measured at the weather station benefited the most. On the other hand, farmers who were located in more distant locations experienced quite different weather patterns and were thus at a distinct disadvantage. For crop micro-insurances the covariant risk is particularly high as well, because many farmers are simultaneously affected. This endangers the capitalisation viability of service providers (Intellecap 2010). Also, in the case of NASFAM the overall decision and logistical power was bundled in this one association, which alleviates transaction costs but resulted in some discontent and a lack of institutional trust among farmers, thus impeding the scheme's potential for scaling up. Causes, among others, were low price setting for crops and lacking competencies to protect against additional agriculture-related risks like diseases and pests. Finally, referring to current climate projections for Malawi, weather-related stress factors are going to increase in the future, which will probably be reflected in increased premiums and expenditures for smallholders, adding to the challenge of creating affordability (Suarez and Linnerooth-Bayer 2010).

Thus, although providing a valuable approach for CBA, micro-insurances still involve a lot of challenges which heavily constrain their potential for scaling up. General difficulties are so far related to weak institutional frameworks, lacking trust in service providers, the insufficient estimations of risks, adverse selection and moral hazards. Particularly in the context of climate change, insurance products are projected to become more expensive in the future, which will not provide a sustainable solution for vulnerable communities in the long term (Suarez and Linnerooth-Bayer 2011).

The role of the private sector in community-based adaptation

Given that there is limited evidence of private sector involvement in climate adaptation at the community level, this section will explore the potential and scope for private sector involvement, using an example. However,

> Sooner or later, all businesses will have to climate-proof their operations, from supply chain to point of sale, from place of production to place of investment. Adaptation will be imperative if businesses want to avoid climate change impacts driving them out of business.
>
> (UNFCCC 2011, p. 1)

Changing climate patterns endanger businesses in various ways, for instance by affecting their core operations and extended value chains involving local communities that compose a company's workforce, suppliers and customers. Although a higher awareness for the necessity to adapt is generated among managers, essential financial resources, capacities, knowledge and institutional frameworks for implementing effective CBA measures are significantly lacking and responsive actions of the required extent are often not executed by the private sector (UN Global Compact 2011).

Defining mutually beneficial objectives for the private sector and CBA has so far only been explored to a limited extent, with most attention currently focusing on how the private sector can contribute to adaptation finance in general, such as generating new finance and helping to fill the massive deficit in available funds for adaptation. Despite acknowledging this necessity, one should not dismiss other valuable contributions the private sector can make for scaling up CBA, which includes technology transfer, capacity building and, in particular, designing, manufacturing and distributing product innovations that may reduce the vulnerability of individuals and communities to climate change, like water harvesting facilities and new plant varieties, as well as providing risk management tools such as insurance (Persson *et al.* 2009).

Scales of private sector players: small to large

The role of the private sector can be further explored by elaborating on the various types and scales of private players that influence community adaptation decisions.

Small informal private players include informal traders in food, fuel wood, transport operators (bicycles, boats, carts) and repairers of equipment. They are very important for the functioning of the local economy.

Small formal operators are registered shop operators, transport operators and providers of communication services that usually link local communities with the outside world, enabling communities to access markets, inputs and services that are not locally available. Local shop operators, for instance, are critical for making goods such as drought-tolerant seed varieties available locally.

Large formal operators could be in the form of outlets of large companies operating in communities – e.g. mobile phone companies, supermarkets or production functions of large companies operating at the community level, wholly or partially using local resources such as land, labour, water and others, thereby interacting with local communities.

The larger the company size the generally stronger their financial strength and capacities to develop and implement CBA in developing countries. Locally run small and medium-sized enterprises (SMEs) are more vulnerable due to their limited resources to cope with and adapt to changing climatic patterns, which endangers their businesses and thus sources of livelihood (Acclimatise and Synergy 2008). This is particularly severe for SMEs that are fundamentally important for creating new employment opportunities, be they formal or informal, at the community level, which contributes to poverty reduction and overall economic growth in emerging economies (CSR 2011).

How the private sector values adaptation

Factors that actually drive a company's strategic decision if and in what way it implements adaptation measures at the community level can be framed around six determining factors, which differ in their importance and extent according to individual business objectives and capacities:

1 Immediate impacts. Companies would only make adaptation an investment priority if the degree of climate risk exposure is immediate by affecting a business's core operations, assets and resources (PWC 2010).
2 Indirect impacts. If the risk is indirect and therefore affects the wider business environment like supply chains, infrastructure, markets and the competition for resources, the awareness for climate change impacts and therefore commitment for adaptation is lower, particularly among principal corporate decision-makers, namely senior board members (PWC 2010). In 2009 the Carbon Disclosure Project survey revealed that 75 per cent of the interviewed managers recognised the negative impacts caused by climate change, whereas only 40 per cent have exercised some kind of business risk estimations, of which 20 per cent have finally implemented adaptation strategies (Agrawala *et al.* 2011). But even if risk appraisals are undertaken, they are often only based on current scenarios and extreme events and thus do not take future changes and ongoing climate variability into account, which increases the risk for maladaptation (Agrawala *et al.* 2011).
3 New products and services. Climate change may constrain the ability of some private sector players to maximise their profits while it may also create new business opportunities for others. For instance, private companies may benefit from buying grain from regions with a good harvest and resale to regions with poor harvests to make a profit. On the other hand, climatic conditions that destroy road infrastructure may work against the private players in the transport sector, which will eventually affect local communities' supply of inputs and access to markets. Emerging commercial opportunities to penetrate new markets for adaptation products and services imply high-profit generation potentials. Exemplary sectors include telecommunications, information technology, food, environmental services, consulting, energy, construction, engineering, financial services, transport, logistics, chemicals, mining and water (Coleman 2011).
4 Regulatory framework. Corporate decisions are heavily shaped by institutional and regulatory frameworks established by the public sector, but these are currently lacking for adaptation (PWC 2010). Stronger policy guidance would generate higher investment incentives and potentially shift business planning horizons, despite the existing climate uncertainty (Agrawala *et al.* 2011). On the other hand, private actors also have the power to shape regulations on adaptation, so activities can indeed be initiated from both sides (UNECA 2010).
5 Climate information. The insufficient reliability of existing climate data aggravates the challenges of non-existent guidance for corporate investments, which consequently often results in inaction (PWC 2010). In this respect, regulatory frameworks initiated by the public sector should also serve as a risk buffer for private investors and leverage them via grants,

guarantees, preferential conditions and innovative financing instruments, which are not widely available yet (UNECA 2010).

6 Corporate social responsibility. Gaining reputational benefits by responding to augmented consumer demands for businesses to become more socially responsible composes additional value, though it would not be a decisive factor for strategic investment by itself. Besides, adaptation as opposed to mitigation measures do not deliver the same favourable publicity since it is more complicated to communicate and benefits are mostly not publicly enjoyed. In fact, businesses may also be reluctant to extensively inform about their adaptation activities since they may compose a competitive advantage (Agrawala *et al.* 2011).

'Community risks are business risks' (UN Global Compact 2011, p. 5). Businesses as well as communities will be adversely affected by the increasing resource scarcity, and thus rivalry induced by climate change will potentially create conflicts and lead to consequences such as production delays, asset destruction, profit losses, as well as costly lawsuits (Acclimatise and Synergy 2008). Additionally, supporting local communities' adaptive capacities through climate change adaptation strengthens a company's licence to operate in the region and improves corporate reputation. Thus, 'the rational self-interest of businesses should be a major driver of adaptation actions' (Agrawala *et al.* 2011, p. 9).

Case study: how Ericsson is enabling access to weather and climate services in Uganda

Sub-Saharan Africa is particularly prone to the adverse climate impacts of floods and droughts that severely endanger local livelihoods and sustainable community development. Climate-induced casualties amount to over 300,000 people every year, while 325 million more are affected. Economic damages add up to over $125 billion (business.un.org 2012).

The continent has witnessed a tremendous increase in mobile phone subscriptions, with an annual growth rate of over 40 per cent, which is estimated to soon become twice as much.

In 2009 the United Nations World Meteorological Organization (WMO), the private telecommunications equipment and service provider Ericsson and the National Meteorological Services established a public–private partnership in Uganda to improve climate observations and facilitate the dissemination of solid weather forecast and disaster warnings through text messaging on mobile phones among fishermen, farmers and small entrepreneurs. With this technological innovation provided by Ericsson, communities are not only enabled to manage their risks more effectively, but also to benefit from economic opportunities in terms of making more informed decisions based on the improved availability of weather data.

After a successful pilot run with fishermen at Lake Victoria in Uganda, it was concluded that the programme has strengthened community livelihoods in terms of human health and food security and also potentially saved lives. Next steps for the project involve extending the mobile climate service to equally weather-dependent smallholder farmers and helping them to adapt their farming schedules and practices according to the weather information provided (business.un.org 2012).

Governments have to remain the major facilitator for adaptation activities through establishing guiding policy frameworks and creating incentivising investment environments. Although more public finances need to be allocated to CBA activities in the future, climate change adaptation at the community level requires enhanced resource inputs from private sector actors. The highest prosperity impacts for both businesses and communities will potentially be generated by the emerging local demands for adaptation products and services, which will autonomously scale-up CBA. The earlier that private actors get engaged in adaptation as first movers, the more they can benefit from the rewards and establish a respective position among stakeholders. However, profit-driven companies are not development agents but need to ensure the commercial viability of their operations, which does not always coincide with the adaptation requirements of poor and vulnerable communities. The public sector needs to fill this gap with existing resources or by enforcing private sector contribution through levying taxes and charges on corporations (SIDA 2009).

However, the existing lack of reliable climate information, particularly regarding most affected sectors and locations, as well as the weak policy framework on adaptation, do not create a robust investment environment for businesses. Therefore, urgent public sector action is required in the form of formulating incentivising regulations to facilitate adaptation spending and respective product development, as well as expanding climate research and distribution of information.

For better defining and achieving mutually beneficial developmental outcomes for communities and businesses, cooperation among the private and public sector, with development agents, research institutes and local communities, needs to be enhanced through the formation of partnerships. Not only to overcome the existing gap between the worlds of climate science and business and the reciprocal resource restrictions, but also learning from local experiences would enable better understanding among the diverse stakeholders and potentially result in more robust adaptation decisions and interventions (Agrawala *et al.* 2011; PWC 2010).

Natural resource-based incentives for CBA

Natural resources are a direct livelihood asset base for most vulnerable communities, providing local communities and society at large with several benefits, including food provision, natural shoreline protection against storms and floods, water quality maintenance, support of tourism and other cultural and spiritual benefits and maintenance of the basic global life support systems (UNEP 2011). The regulating and provisioning roles indicate their importance for climate change mitigation and adaptation. The Millennium Ecosystem Assessment (2005) established the linkage between ecosystem services and human well-being, specifically highlighting the role of ecosystem services in regulating climate, floods, diseases and water purification, as well as in provisioning. The inter-connectivity between ecological, socio-cultural, economic and institutional structures (UNEP 2011) suggests an economic dimension to natural resource-based adaptation to climate change. Natural resources provide communities with several direct and indirect benefits which can be expressed in monetary and non-monetary terms, and these benefits could contribute to community and household adaptive capacity.

Community involvement in natural resources management has largely been driven by incentives associated with the sustainable management of those resources. The incentives are derived from natural resource-based enterprises and payments for ecosystem services. The evolution of community-based natural resources management (CBNRM), for example, illustrates the potential role of incentives for CBA (Chishakwe *et al.* 2012). Payments for ecosystem services also show the role of incentive mechanisms at the community level to sustainably manage natural resources. Chishakwe *et al.* (2012) identify the potentially close linkages between CBNRM and CBA through the use of incentives. While CBNRM was initially seen primarily as a conservation approach, it emerged that it also had rural development co-benefits. It is now viewed as an institutional or organisational development programme whereby natural resources are utilised to economically empower local people. This has proved to be an incentive to organise communities around democratic and managerial principles. The spread of CBNRM initiatives in Southern Africa and other regions of the world, as well as that of payments for ecosystem services (PES) provides a strong basis for scaling up incentive-based CBA built upon the natural resource base that supports most vulnerable communities. In Zimbabwe, the Communal Areas Management Programme for Indigenous Resources (CAMPFIRE) increased in scale from two districts in 1989 to 12 in 1993, with the benefiting population rising from 55,000 to 480,000 in the same period (see Table 4.3).

Incentives exist in various forms, including: remunerative incentives where action is motivated by material reward (financial or non-financial); moral

Table 4.3 The rate of growth and scale of the CAMPFIRE programme in Zimbabwe

Year	*Number of districts with appropriate authority status*	*Number of wards receiving benefits*	*Number of households benefiting*	*Population benefiting*
1989	2	16	7,861	55,000
1990	12	30	22,040	155,000
1991	12	66	51,938	365,000
1992	12	70	70,610	495,000
1993	12	70	68,798	480,000

Source: adapted from Chishakwe *et al.* (2012).

incentives where action is motivated by moral satisfaction; coercive incentives where action is motivated by fear of punishment; and natural incentives where action is motivated by subjective human interests (Sheffrin and Sullivan 2003). In fact, the analysis of CBNRM projects in Southern Africa shows that non-cash benefits of CBNRM have proved to be an important driver for conservation and rural development in Zimbabwe, Namibia and Botswana, with established conservancies in Namibia achieving higher welfare levels than those in their infancy, but not necessarily high cash incomes for households, but due to the importance of in-kind benefits such as game meat. In Botswana, communities place higher value on the fact that CBNRM empowers them and gives them choices for development and livelihood improvement (Chishakwe *et al.* 2012). The cash and non-cash benefits from incentive-based natural resources management have potential to contribute to climate change adaptation at the community level.

Another important point is the role of natural resource-based economic opportunities in upscaling CBA. PES are an increasingly popular innovative market-based instrument approach to conservation that has been applied increasingly often in both developed and developing countries to translate external, non-market environmental services into financial incentives for local actors to provide environmental services. Community-based natural resource management has also been spreading through South–South cooperation, with the Southern African model of community ecotourism being applied in Asia (www.iied.org/sustainable-markets/key-issues/environmental-economics/sustainable-tourism-srepok-wilderness-area-ca).

Conclusions

Economics is a very useful tool for adaptation decision-making, and economic approaches can be used to facilitate or enhance adaptation. In the context of CBA, the role of economic tools and approaches has to be carefully evaluated to take into account the way in which decisions are made at the local level. Economic efficiency considerations such as cost–benefit criteria applied in isolation may fail to reflect the multiple goals that communities pursue together with climate adaptation pursuits. The way in which communities value adaptation inputs and outputs may differ from their valuations using market prices. Some adaptation actions may not be autonomously scaled up even when economic criteria are favourable because of the assumptions made about valuation of costs and benefits. Market approaches, the involvement of the private sector and incentive mechanisms for natural resource-based approaches will not enhance scaling up of adaptation to full potential because of imperfections in markets. Often, communities do not have full information, have poor infrastructure leading to loss of competitiveness on the market or do not meet the rigorous standards of, in particular, export markets. These realities suggest that economic approaches should be used in combination with other practical considerations.

CBA is an emerging field, and the evidence on application of economics in the scaling up of CBA is still limited. Much of the evidence is coming from other fields such as community-based natural resource management, agricultural enterprises and community development, where the data and analyses available cover long periods of time. These, however, provide a good basis for informing CBA, and suggest the areas for monitoring its evolution. Only a few aspects of economics have been highlighted here, leaving room to include other dimensions of economics to CBA. This chapter has deliberately not attempted to pursue a theoretical discussion of the economics of CBA. Instead, it has highlighted the different perspectives, paving the way for more specialised treatment of the subject as the field evolves.

Note

1 www.adaptasertao.net.

References

Acclimatise and Synergy, 2008. *Climate Finance, Business and Community: The Benefits of Co-operation on Adaptation*. Discussion Paper.

Agrawala, S., M. Carraro, N. Kingsmill, E. Lanzi, M. Mullan and G. Prudent-Richard, 2011. *Private Sector Engagement in Adaptation to Climate Change: Approaches to Managing Climate Risks*. OECD Publishing.

Alcala, A.C., A.A. Bucol and P. Nillos-Kleiven, 2008. *Directory of Marine Reserves in the Visayas, Philippines*. Foundation for the Philippine Environment and the Siliman University – Angelo King Center for Research and Environmental Management (SUAKCREM).

business.un.org, 2012. *Mobile Weather Alerts: Enabling Access to Weather and Climate Services in Africa*. Retrieved 28 January 2012 from http://business.un.org/en/documents/9302.

Chambwera, M. and J. Stage, 2010. *Climate Change Adaptation in Developing Countries: Issues and Perspectives for Economic analysis*. International Institute for Environment and Development.

Chishakwe, N., L. Murray and M. Chambwera, 2012. *Building Climate Change Adaptation on Community Experiences: Lessons from Community-Based Natural Resource Management in Southern Africa*. International Institute for Environment and Development.

Coleman, H., 2011. *Business Partnership to Promote Resilience and Environmental Preparedness*. Retrieved 2 January 2012 from www.oxfamamerica.org/articles/business-partnership-to-promote-resilience-and-environmental-preparedness-forms.

CSR, 2011. *Climate Change Adaptation: Engaging Business in Asia. Main report.* Retrieved 1 January 2012 from www.unisdr.org/files/20777_sida21.pdf.

Fairtrade Foundation, 2011. *Annual Report and Financial Statements for the End 31 December 2010*. www.fairtrade.org.uk/includes/documents/cm_docs/2011/2/1_2010%20Report%20and%20Accounts%20FINAL.PDF.

Intellecap, 2010. *Opportunities for Private Sector Engagement in Urban Climate Change Resilience Building*. Retrieved 28 January 2012 from http://unfccc.int/files/adaptation/application/pdf/intellecap_furtherinfo_250311.pdf.

Millennium Ecosystem Assessment, 2005. *Ecosystems and Human Well-being: A Synthesis*. Island Press.

Persson, A., R.J.T. Klein, C.K. Siebert, A. Atteridge, B. Müller, J. Hoffmaister, M. Lazarus and T. Takama, 2009. *Adaptation Finance under a Copenhagen Agreed Outcome*. Stockholm Environment Institute.

PWC (PricewaterhouseCoopers LLP), 2010. *Business Leadership on Climate Change Adaptation: Encouraging Engagement and Action.* Retrieved 30 December 2011 from www.pwcwebcast.co.uk/encouraging-engagement-and-action-full-report.pdf.

Sheffrin, S.M. and A. Sullivan, 2003. *Economics: Principles in Action.* Pearson Prentice Hall.

SIDA (Swedish International Development Cooperation Agency), 2009. *Making Climate your Business: Private Sector Adaptation in Southeast Asia.* Retrieved 1 January 2012 from http://pdf.wri.org/making_climate_your_business.pdf.

Stern, N., 2006. *The Economics of Climate Change: The Stern Review.* Cambridge University Press.

Suarez, P. and J. Linnerooth-Bayer, 2010. Micro-insurance for local adaptation. *Wiley Interdisciplinary Reviews: Climate Change*, 1: 271–278.

Suarez, P. and J. Linnerooth-Bayer, 2011. *Global Assessment Report on Disaster Risk Reduction 2011. Insurance Related Instruments for Disaster Risk Reduction.* Retrieved 23 January 2012 from www.preventionweb.net/english/hyogo/gar/2011/en/bgdocs/Suarez_&_Linnerooth-Bayer_2011.pdf.

UN Global Compact, 2011. *Adapting for a Green Economy: Companies, Communities and Climate Change. A Caring for Climate Report.* Retrieved 2 January 2012 from www.unglobalcompact.org/docs/issues_doc/Environment/climate/C4C_Report_Adapting_for_Green_Economy.pdf.

UNECA (United Nations Economic Commission for Africa), 2010. *Seventh African Development Forum VII: Acting on Climate Change for Sustainable Development in Africa. Private Sector Response to Climate Change.* Retrieved 2 January 2012 from www.uneca.org/adfvii/documents/IssuePaper11Private-sector-response-to-climate-change.pdf.

UNEP (United Nations Environment Programme), 2011. *Towards a Green Economy: Pathways to Sustainable Development and Poverty Eradication*, www.unep.org/green-economy.

UNFCCC (United Nations Framework Convention on Climate Change), 2011. *The Private Sector Initiative of the Nairobi Work Programme on Impacts, Vulnerability and Adaptation to Climate Change.* Retrieved 23 December 2011 from http://unfccc.int/files/adaptation/nairobi_work_programme/joining _the_nwp/application/pdf/nwp_psi_facsheet_may2011.pdf.

5 Seeking sustainable financing mechanisms for upscaling community-based adaptation

Fumiko Fukuoka, Charles Nyandiga, Paul Mitchell, Achala Chandani, James Hardcastle, Andrew Adwera, Aminul Islam, Sarwat Chowdhury, Albon Ishoda and Leena Wokeck

Because adaptation to climate change is cross-sectoral in nature and affects key development areas, it has considerable impact on the lives and livelihoods of natural resource-dependent local communities. Therefore, investing in community-based adaptation (CBA) is increasingly recognised as a strategic approach for providing vulnerable populations with ground-level assistance while maximising development impacts in a holistic manner (UNFCCC Subsidiary Body for Scientific and Technological Advice, 2010, INF.7).

Globally, annual investments in the range of \$49–171 billion are required to finance an adequate level of climate change adaptation. In developing countries, where climate change impacts are most severe, annual investments in the range of \$26–87 billion are needed to cope with the burdens of climate change (UNFCCC Subsidiary Body for Scientific and Technological Advice, 2010, INF.7).

While specific attention can be given to demand-driven CBA initiatives, it is important that CBA be recognised as an integral aspect of investment in the scope of broader development assistance. Therefore, CBA must access, combine and sequence funding opportunities that are catalytic, innovative and diversified in nature, and can support ground-level implementation.

As the global architecture for funding climate change adaptation progresses it is necessary to dissect the opportunities and challenges of directing finance towards CBA. This chapter emphasises the importance of CBA by examining global financing trends and incentivisation of local opportunities to exploit limited public and private resources in order to leverage private and national finance instruments for enhanced realisation of CBA. In addition, we will analyse challenges and difficulties, define observations and recommend a way forward to create a sustainable finance mechanism for upscaling CBA investments.

Global trends for multilateral financing

In recent years considerable attention has been devoted to the need for public finance to play a larger role in providing aid to developing countries and communities vulnerable to the impacts of climate change. For example, in 2010 developed countries collectively committed to raise $30 billion, vowing to allocate half of that amount to climate change adaptation for developing countries and communities. Parties to the UNFCCC have committed to continue funding adaptation in the coming years and have also undertaken to mobilise $100 billion in annual climate change assistance by 2020. While these commitments are welcomed, they are far below estimated needs (Stern 2006; World Bank 2006; Oxfam International 2007; UNDP 2007; Parry *et al.* 2009).

Currently, global financial support flows through two main channels: multilateral funds and bilateral donor programmes. While CBA activities receive funds from both sources, there remains significant potential for the scale-up of investments.

Multilateral finance and CBA

Increased multilateral finance for CBA would allow vulnerable communities to better address the devastating impacts of climate change. Therefore, the provision of adequate financial resources is a key element in addressing the challenges of adapting to the adverse effects of climate change within the global climate change regime. Under the UNFCCC and the Kyoto Protocol (KP), developed countries commit themselves to supplying financial resources for the implementation of adaptation activities in developing countries (Article 4.4).

Of the three main funds under the global climate change regime, the Least Developed Countries Fund (LDCF) is designed to support a work programme for assisting least developed countries (LDCs) with, inter alia, the preparation and implementation of national adaptation programmes of action (NAPAs). The Special Climate Change Fund (SCCF) also finances, inter alia, projects related to adaptation, while the Adaptation Fund (AF) finances concrete adaptation projects and programmes in developing countries that are parties to the KP, and are particularly vulnerable to the adverse effects of climate change. Thus, UNFCCC provisions related to the three main funds, or any of the related decisions on financing adaptation, are not rooted in CBA.

There are unexplored opportunities for UNFCCC parties to address CBA directly within the global process. For example, decision 10/CP.7 introduces guidelines and criteria for funding under UNFCCC or the Nairobi Work Programme. Despite introducing nine areas of work, the decision does not refer to CBA. Similarly, the guidelines for national communications have no reference to CBA. Nevertheless, a close analysis of UNFCCC provisions on adaptation financing support the argument that opportunities do exist for financing CBA activities under the global regime. For example, the direct access modality of AF may help developing countries to address CBA activities more easily through national implementing entities. AF is the first multilateral climate change fund that implements an institutional structure for direct access on a regular basis (decision 1/CMP.3). Under the existing funding structure of the financial mechanism of the UNFCCC, eligible countries can access funds only indirectly, through the implementing and executing agencies of the Global Environment Facility (GEF). The direct access modality would facilitate approaches for country-driven project identification, and provide improved operations.

According to key principles still being discussed under the new Green Climate Fund (GCF) and the Cancun Adaptation Framework, parties do have the opportunity for allowing direct CBA financing in the future. However, many concerns related to the GCF remain unresolved. For example, it is unclear whether CBA will be a key funding principle under GCF. According to the Cancun Agreement, only 'countries most vulnerable to the effects of climate change are to be prioritized in the allocation of finance', and nothing in the language refers to communities (UNFCC Ad hoc Working Group on Long-Term Cooperative Action under the Convention, 2010, INF.1).

However, there exist other opportunities for financing CBA activities under the global regime. For example, because CBA elements have been recognised under the Cancun Adaptation Framework, the Adaptation Committee can be shaped to ensure CBA is taken into account – as can the newly introduced National Adaptation Plans. In addition, with Fast Start Finance (FSF) donor countries can show commitments to CBA. For example, Germany has allocated some of its FSF for CBA, and it is expected that, after the Cancun Agreements, community-level adaptation will feature more prominently. How this will be reflected is not yet clear. One possibility is the introduction of a dedicated funding mechanism for CBA. This would provide communities with predictable long-term financing and direct access to funds. However, there are challenges to address before introducing a dedicated funding mechanism for CBA. These problems are outlined below.

- Limited funding. In terms of actual funding, the LDCF and SCCF were established in 2002 and 2006, respectively (GEF, d.u.). Meanwhile, the AF only signed its first project agreement in November 2010. To date, the LDCF has approved only $346 million for 74 projects and one programme in 44 countries (as of June 2012), and the SCCF has mobilised $162.24 million for 39 projects and three programmes in developing countries (as of June 2012). With regard to AF, the amount of funds available depends on the generation and monetisation of Certified Emission Reductions (CERs) from the Cleaner Development Mechanism (CDM), the market price and donations to the AF. Using the medium case assumptions, as of June 2012 the AF has funded 25 countries, amounting to $41 million out of an available $166.
- The lack of a clear definition of the term 'vulnerability'. Vulnerability can be used as a criterion to decide allocation of global financial resources among countries or regions. At the national level, such vulnerability

assessments can play an important role in guiding national policy action and prioritising the flow of financial resources to its communities. Vulnerability assessments consider economic, social and ecological systems and needs and inform concrete actions on the ground. However, the definition of 'vulnerability' remains broad and often undefined at the global level; there is no agreed metrics or set of indicators for vulnerability assessments. In some cases, the possibility of using vulnerability as a criterion for allocating global finance is considered. For example, the possibility for the AF Board to allocate funding based on a vulnerability-oriented funding approach was considered. However, this still remains largely unspecified and unused as the global level funding under the UNFCCC regime is based on the principle of a country-driven approach. Although the Strategic Priorities, Policies and Guidelines of the AF mentions that 'special attention shall be given by eligible Parties to the particular needs of the most vulnerable communities' (1/CMP.4 Annex IV) the strategic guidelines do not include vulnerability as a criterion in the assessment of project and programme proposals (AFB/B.7/13/Rev.1 Annex III). The AF encourages countries to consult information provided by the Intergovernmental Panel on Climate Change (IPCC) and information and reports generated under the Nairobi Work Programme on Impacts, Vulnerability and Adaptation to Climate Change (AFB/B.6/5) in relation to vulnerability; the information and data available in these reports are often not adequate to map the vulnerability situation globally. Above all, they do not allow prioritisation of activities that are geared towards CBA.

- Lack of institutional capacities and effectiveness. Institutional architecture provides the backbone for transfer of financial resources from the global level and effective support to communities. The need for institutional arrangements in order to respond to climate change impacts has been recognised throughout the UNFCCC. However, the lack of institutional capacity that effectively facilitates the funding process remains a central challenge the developing countries are facing in accessing funding from the global level and effectively allocating it at national, local and community levels. For example, the UNFCCC AF, the first multilateral climate change fund that implements an institutional architecture for direct access, has allowed countries to nominate their National Implementing Entities that can directly access funds. According to the Operational Policies and Guidelines for Parties to Access Resources from the Adaptation Fund (AFB/B.7/13/Rev.1 Annex III), such institution can access the adaptation fund for financing concrete adaptation activities in the form of projects or programmes aimed at addressing the adverse impacts of and risks posed by climate change that communities, countries and sectors are facing. Developing countries have been asking for such direct access modality for many years. The direct access modality represents a breakthrough in the international funding mechanism and aims at simplifying and accelerating the way by which the Fund disburses its resources. However, to date, only eight countries have managed to accredit their National Implementing Entities and less than 10 per cent of funding has been disbursed through these National Implementing Entities. Such lack of institutional capacity at the national level hinders the progress made on effective funding for CBA.

- Potential to isolate communities from other sources of finance and reduce the total funds available for CBA. A dedicated funding mechanism for CBA may or may not result in isolating communities from other sources of funding. On the one hand, support needed for CBA may include addressing all environmental risks, including disaster risks and health problems, in order to address vulnerabilities of the poor communities. On the other hand, funding may be required for local development that may only implicitly increase resilience to likely impacts of climate change. Such vague areas may not be eligible for adaptation financing from the global level. However, when there is a clear component of climate change involved in a project, such projects are fundamental under current multilateral funding systems. For example, analysis of the LDCF funding portfolio suggests that agriculture and water management have been, by far, the most important project components funded to date under the LDCF window. Disaster preparedness, coastal zone management, health and infrastructure have also been targeted in the GEF adaptation portfolio.

Bilateral principles and links to national-level financing

Though bilateral donors are scaling up support for adaptation, the focus has been on national and regional activities and global funds. In order to leverage a larger share of available and planned climate finance for CBA, practitioners should take the following actions.

Seek opportunities to generate economies of scale

Donors are managing increasingly large and complex funding streams for climate change adaptation, working across a wide variety of sectors in many countries. Often, donors perceive managing a number of small projects an unnecessary burden. In this context, CBA practitioners should seek opportunities to generate economies of scale by creating inter-linkages among communities facing similar challenges. International non-governmental organisations (NGOs) and multilateral agencies have an advantage in this sphere due to their ability to handle larger funding flows. However, engagement with community-based organisations and national NGOs is of paramount importance to ensure activities are locally owned and CBA investments address community priorities. Donors – including Australia, Germany, Norway, the United Kingdom and the United States – are running competitive grant windows for international NGOs implementing CBA activities with local partners across multiple communities and countries. The benefits of this approach include ease of management and wider reach for donors, as well as the ability for NGOs to generate cross-learning, South–South cooperation and an overall increase in funding flows to CBA.

Ensure CBA activities contribute to national development goals

To be effective, bilateral aid programmes need to align with the priorities of partner governments and support the implementation of national development strategies. Given the potential for unwanted overlap and duplication between adaptation funding and official development assistance, donors are increasingly aiming to integrate adaptation into their broader development programmes. In this context, CBA practitioners should ensure proposed CBA activities have a key objective of helping local communities contribute to broader national-level development goals. Stronger linkages between CBA and national policies also create opportunities for communities to influence national climate change policy by showing what works on the ground. An example of this interaction and its impacts is the integration of CBA into the National Climate Change Roadmap 2010 in the Republic of the Marshall Islands (RMI).

Engage at the regional and global levels

The principle of integration at the national level is also valid at regional and global levels. While deep engagement in individual countries is important, broad engagement across countries increases effectiveness and promotes innovation by linking different communities and highlighting examples of best practice. International organisations and NGOs are already creating mechanisms to link local-level action into a global community of practice through the Global Initiative on Community Based Adaptation. In other words, global and regional CBA mechanisms can leverage off existing mechanisms to minimise proliferation of management structures and ease administrative burdens on communities. For example, Australian funding enables the GEF Small Grants Program (SGP) to expand a ten-country CBA programme to cover 42 more countries.

Adherence to these three principles should secure additional bilateral donor interest in CBA and enable practitioners to generate a body of knowledge on what works on the ground to feed into national policies and international processes.

Incentivising private investments for CBA

There is a strong link between economic development, business growth and climate change adaptation. Business plays a crucial role in driving economic development and building adaptive capacity in developing countries and vulnerable communities. Building on core competencies, business can offer skills that can be used to adapt to climate change.

The role of the private sector in climate change adaptation is becoming more prominent, with private finance regarded as critical to meeting adaptation goals. However, strategies to effectively incentivise these investments remain vague. Improved consultation and engagement processes will be needed to identify opportunities for allocating public money in ways that effectively leverage

Box 5.1 Case study: enabling CBA on a national scale in the Republic of Marshall Islands

The RMI consists of over 125 islands and remote atoll systems spread over a large area of the Pacific Ocean. A number of enabling conditions have led to the scaling up of CBA into national policy, such as the RMI National Climate Change Roadmap 2010. Some of these enabling conditions are outlined below.

- Acknowledgement of local context. Apart from the more urbanised centres of Majuro and Ebeye, most atolls are remote and support small communities reliant on traditional subsistence lifestyles, which are heavily impacted by the pressures of climate change. It is acknowledged that any national plan for climate change adaptation must embrace CBA to ensure these outer atolls are considered.
- External drivers. A presidential commitment made in 2006 to the Micronesia Challenge, which calls for 20 per cent of near-shore coastal areas and 30 per cent of terrestrial areas to be protected and effectively managed by 2020, has created a pathway for resources and technical assistance to flow to remote communities.
- Multi-sector support. In 2007, the Coastal and Marine Action Council (CMAC) was formed to help integrate support to outer atolls. CMAC comprises representatives from key departments and agencies within government, as well as national NGOs and civil society. CMAC convened a number of CBA demonstration projects and activities that explored how to integrate lessons into policies and approaches.
- Using demonstrations to inform planning processes. In the remote Namdrik Atoll, early consultations by the Marshall Islands Conservation Society (MICS) on behalf of CMAC raised a number of concerns related to ecosystem integrity and the impacts of climate change on health, water, waste management, infrastructure, livelihoods and food and fish availability. With domestic and international project support, Namdrik produced an Atoll Development Plan that fully integrates CBA, and is endorsed by the informal Iroji traditional leaders. Lessons learned from Namdrik helped inform the Reimaanlok ('Way Forward') – a national guidance document on community-based natural resource management and climate change adaptation, which is a component of the National Climate Change Roadmap 2010.
- Linking CBA and economics. Based on their ground-level experience with start-up and recurrent support, groups like MICS have estimated the cost of implementing CBA. The Micronesia Challenge sustainable finance plan for the Marshall Islands is being developed to meet the required demand for CBA, starting with an endowment fund managed by the Micronesia Conservation Trust. This will help provide the shortfall of funds required to implement CBA beyond national resource flows.

As a result of these enabling conditions at local, national and regional scales, multiple sectors are increasingly able to integrate CBA into their strategies and respond more efficiently to community demands for climate change adaptation measures on the outer atolls.

additional private funding. The public sector will need to create clear framework conditions that incentivise the private sector to allocate resources to meet adaptation needs.

Companies with sophisticated corporate social responsibility (CSR) strategies tend to understand and seek to capitalise on non-financial returns on investments. The increasing preparedness of businesses to integrate non-financial calculations into their balance sheets is an important lever to encourage effective engagement with CBA

In addition to positive incentives, there may be a need for regulatory mechanisms that mandate business planning for climate change impacts (e.g. requiring businesses to invest in disaster response strategies and business continuity planning, and to protect privately owned assets, including those of public significance through privately financed insurance schemes).

The most important driver for business to act on climate change is the need to safeguard operations, assets and value chains. Enhanced economic viability of progressive adaptation will require the implementation of certain strategies:

- Set clear policy directions to provide businesses with greater certainty for business planning. For example, government leadership is required to ensure the integration of adaptation concerns into building regulations, standards and design criteria.
- Provide adequate information services on the impacts of climate change on specific geographical areas and industry sectors.
- Improve the allocation of public efforts and resources to incentivise private investments. This requires improved alignment of public funding mechanisms, clear directions, priorities, allocation criteria and comprehensive impact measurement systems.
- More effectively engage business in public and development processes for climate change adaptation. This applies to international, national and local planning.
- Contribute to capacity building in the private sector, in particular for small, medium and micro-enterprises. Multidisciplinary sector initiatives can help advance the awareness and availability of response options in industry sectors particularly vulnerable to climate change, benefiting both private and public interests.

In addition, approaches to climate change adaptation must be linked to a broader development agenda and prioritise efforts to build resilience of vulnerable communities. Business can play an important role in improving the economic, social and environmental circumstances of communities. Leading businesses have devised strategies to help contribute to sustainable community development at the local level through community investment programmes and inclusive business models.

Strategic community investment (CI) refers to 'voluntary contributions and actions taken by companies to help communities in their areas of operation address their development priorities, and take advantage of opportunities created by private investment – in ways that are sustainable and support business objectives' (IFC, 2010). CI looks beyond finance to achieve greater community impact. It makes effective use of core business expertise and relationships to contribute to community development areas that are selected based on locally identified community needs.

Unfortunately, CI remains limited. Most companies are engaged in corporate philanthropy. In addition, climate change and CBA are seldom addressed systematically in CI strategy and programme development. Much needs to be done to better harness the innovative capacity, resource wealth and implementation efficiency of the private sector for effective response options to development challenges that exacerbate adaptation needs:

- Few businesses have developed climate change adaptation strategies that focus on the link between climate change adaptation and development and tackle the underlying causes of vulnerability. Businesses will need to increasingly recognise the strategic relevance of integrating sustainability and development issues into their business strategies for the development agenda to be effectively promoted as a viable entry point to climate change adaptation.
- Because estimates of adaptation costs are high, businesses are primarily sought out for additional funding. This limits consideration of other, potentially more effective and sustainable, contributions that businesses can make to climate change adaptation, including technical expertise, effective planning and management approaches and innovative capacity to meet changing needs.

Engaging with businesses in partnerships for development has the potential to successfully address underlying drivers of vulnerability and help build the resilience of vulnerable communities. This is illustrated by a mechanism in Kenya where funds transferred from government to grassroots communities incentivises the private sector to contribute to development agendas:

Box 5.2 An integrated approach to bridging funds

In Western Kenya, the Uhai Lake Forum, a community-based organisation (CBO), receives technical assistance from many partners, including the International Institute for Environment and Development (IIED) and the African Centre for Technology Studies (ACTS). ACTS, working with Uhai and other grassroots organisations, has been able to demonstrate how the district development office (DDO) can act as a bridge between the national government, private sector and communities to channel funds from national entities to be disbursed at local levels based on the recommendations of district development committees.

ACTS has also developed new tools, such as videos prepared through participatory process to monitor the use of resources at the community level, providing practical ways for the community to track climate impacts and showing how allocated resources are flowing to target groups.

Positioning of selected typologies of CBA projects

Scrutiny of the past five CBA conferences where new practices and innovations were shared attests to an increased demand for projects related to disaster abatement. While pilot CBA projects have been undertaken in different ecosystems and landscapes by various civil society organisations (CSOs), governments and funding facilities (e.g. GEF and bilateral and multilateral organisations), there has been neither a clear pattern of preference, nor any considerations that funding agencies find attractive. As a result, CBA activities at local levels remain largely ad hoc and reactive.

However, patterns of common project positions have emerged (e.g. planning project locations based on what is practical and vulnerability maps). For example, both the World Bank (WB) marketplace small grants projects and a new programme supported by the Government of Japan, Satoyama (http://satoyama-initiative.org/en/case-studies) exhibit the same positioning trends as locations found in the Google Map layer. These trends are the norm for landscape-based community projects, which often display elements of deliberate or indirect response to the variability of climate change.

A systematic analysis of these global trends will help decision-making parties channel resources to sectors with the highest potential impacts – not necessarily the most vulnerable communities or sites – on livelihoods and rural economies. Such sectoral approaches are increasingly evident in ongoing projects. For example, the UNDP-GEF Strategic Priority on Adaptation (SPA) project shows priorities are being directed to certain thematic sectors. In SPA CBA projects, specific themes have emerged. These themes include capacity building to implement new and affordable technology (e.g. hill slope stabilisation structures and drip irrigation systems) and the management of water resources and land degrading risks. In other words, CBA projects tend to concentrate on risk aversions informed by perceived impending disasters, as illustrated by a case study examining the experience of pooled funding for CBA in Bangladesh.

The trend exhibited by sample project locations in this study (Box 5.3) can also be grouped into projects aimed at increasing adaptive capacity and resilience of communities. Again, while CBA projects are typically sector oriented, the trend toward multifocal projects points to desectoralisation – a manifestation of community prioritisation patterns, which tend to be all-inclusive. This runs counter to secular funding from agencies and global mechanisms, which are designed to respond with little flexibility to specific, sector-based issues. Prioritisation of support areas and themes by funding facilities (e.g. GEF) generally respond to identified global benefits, which in most cases are not of immediate interest to communities. Complications arise with the need to merge global benefits expected by the scientific community with local priorities; especially because CBA projects funding typically range across \$25,000–50,000 over a project lifetime of 18–24 months. While it is generally believed that communities can best manage smaller amounts of resources, local voices are requesting funds in excess of \$200,000 per project over 36 months or more.

Box 5.3 Case study: the Local Disaster Risk Reduction Fund (LDRRF) in Bangladesh

In the evolving field of CBA, disaster risk reduction and climate change adaptation are inextricably related. Therefore, funding mechanisms for upscaling CBA can take lessons learned from successful activities in the field. The LDRRF experience in Bangladesh is a model case for utilising planning tools that can be adopted by other LDCs.

The LDRRF's key objective is to support implementation of risk reduction measures identified by vulnerable communities through community risk assessment. This leads to the creation of Risk Reduction Action Plans, which in turn lead to the sanctioning of small grants. LDRRF provides these grants to NGOs and local governments, who implement risk reduction measures at the community level under the leadership of Disaster Management Committees.

This innovative funding mechanism was established under the Comprehensive Disaster Management Programme, a recently up-scaled multi-donor disaster risk management programme implemented by UNDP in Bangladesh. It includes integration of disaster risk reduction and climate change adaptation, as well as engagement of local governments. So far, the LDRRF has provided funds for 699 small-scale risk reduction interventions across five districts of Bangladesh, and will eventually reach 40 districts vulnerable to climate change. In the LDRRF model, disaster risk reduction includes consideration of climate-induced phenomena (e.g. floods, cyclones, drought), which threaten the lives and livelihoods of millions of Bangladeshis. Institutionalising this integrated process of community risk assessment via further upscaling of methods employed by the LDRRF will provide the opportunity to bridge the gap between scientific information and community experiences, which will trigger the development of more effective Risk Reduction Action Plans.

Emerging funding patterns for CBA projects

As mentioned, resource funding for CBA is increasingly combined with national development actions – a pattern shaped by aid and donor agencies, bilateral organisations, civil society, governments and communities. Governments continue to recognise this trend and seek control of the processes because CBA projects have great potential to influence policies, and are attractive entry points for grassroots mobilisation.

Similarly, funding agencies are demanding that communities provide local infrastructure and basic non-monetised services to reduce the costs of adaptation delivery at local levels. Funding strategies have therefore aimed at penetrating government bureaucracies and exploiting community social networks and economies, thereby increasing local control and decreasing pressures exerted on communal resources. For maximum impact, these strategies require increased mergers and alliances from all sectoral support, which may increase community resilience, leading to decreased external intervention.

Most external support for communities coalesces in the most vulnerable parts of the world and society: sub-Saharan-Africa, where vulnerability is related to livelihood support systems; Latin America, particularly the northern regions inhabited by indigenous communities; small-island developing states, which face threats from sea-level rise; and LDCs with poverty-related vulnerability. Google Map (http://weadapt.org/placemarks/earth#) and WB Development Marketplace (WB, d.u.) are important tools that capture projects and globally georeference their locations – both tools reinforce the trending patterns. At the global scale, Africa is the region most vulnerable to climate change risks, followed by Asia, Latin America, the Caribbean and Pacific and Europe (Maplecroft 2010). It is important to note that though the Google Earth layer is inaccessible to many organisations, and therefore not well populated, if utilised properly it would most likely reaffirm this trend.

In countries where there is an existing portfolio of CBA projects, a strategic approach should be undertaken to replicate or up-scale the existing initiatives, using country vulnerability maps and new lessons complemented by innovations and technologies generated from ongoing initiatives.

Conclusions and recommendations for ways forward

This chapter discussed core global issues with regard to scaling-up CBA. This was done by exploring and mapping the funding architecture, and suggesting possibilities for mainstreaming CBA into national agendas and garnering increased support for CBA in global climate finance architecture.

While future climate change funding flows are expected to approach $100 billion annually by 2020, the general consensus is that this will be insufficient to address impacts. Competition for funding will likely intensify as donors and multilateral funds shift to direct national implementation and budget support. In order to access funding, CBA will need to prove itself as a viable, effective and efficient mechanism for increasing the resilience of poor communities. With a short time frame and limited funding this will prove challenging.

Working in partnership with key actors and players is critical to jointly develop work programmes and fund raising to seek innovative financing strategies that are catalytic in nature – for example, setting up infrastructure and logistics for implementing CBA projects on the ground using modalities (e.g. SGP for delivering community projects) that are already available.

Forging new alliances and facilitating collaboration among national governments, international development organisations and the donor community to ensure synergy and complementarity of effort in supporting the development and accessibility of funds from entities such as GCF will require lobbying by CSOs and CBOs. The Cancun Agreements acknowledged the importance of focusing on the 'urgent and immediate needs' of those who are most vulnerable to climate change (Decision 1/CP.16, paragraph 11). CBA can become a key mechanism through which these needs are met.

Development partners should prioritise the generation and dissemination of knowledge products to up-scale community-based initiatives and support coherent policy development to integrate CBA into national development agendas. In this respect, the links among functioning ecosystems, sustainable livelihoods and resilient communities provides opportunities for businesses to invest in protecting ecosystems, thereby building the adaptive capacity of communities along their value chain. Awareness raising and capacity building, effective communications on successful demonstration projects and policy dialogue will need to be strengthened to promote future uptake.

References

GEF. (2010) *Status Report on the Least Developed Countries for Climate Change*. 1 May.

IFC. (2010) *Strategic Community Investment: A Good Practice Handbook for Companies Doing Business in Emerging Markets*, IFC, Washington, DC. Online, available at: http://ifc.org/ifcext/sustainability.nsf/content/publications_handbook_community investment,GEF, accessed 15 September 2012.

Oxfam International. (2007) *Adapting to Climate Change: What's Needed in Poor Countries, and Who Should Pay?*, Oxfam, Oxford.

Parry, M., Arnell, N., Berry, P., Dodman, D., Fankhauser, S., Hope, C., Kovats, S., Nicholls, R., Satterthwaite, D., Tiffin, R. and Wheeler, T. (2009) *Assessing the Costs of Adaptation to Climate Change: A Review of the UNFCCC and Other Recent Estimates*, International Institute for Environment and Development and Grantham Institute for Climate Change, London.

Stern, N. (2006) *Stern Review: Economics of Climate Change*, Cambridge University Press, Cambridge.

UNDP. (2007) *Human Development Report 2007/2008: Fighting Climate Change – Human Solidarity in a Divided World*, Palgrave Macmillan, New York.

UNFCCC Ad hoc Working Group on Long-Term Cooperative Action under the Convention. (2010) *In-session Draft Texts and Notes by the Facilitators Prepared at the Twelfth Session of the Ad hoc Working Group on Long-term Cooperative Action under the Convention, Note by the Secretariat*, 29 October.

UNFCCC Subsidiary Body for Scientific and Technological Advice. (2010) *Progress Made in Implementing Activities under the Nairobi Work Programme on Impacts, Vulnerability and Adaptation to Climate Change*, 4 November.

World Bank. (2006) *Investment Framework for Clean Energy and Development: A Progress Report*, World Bank, Washington, DC.

6 How is community-based adaptation 'scaled up' in environmental risk assessment?

Lessons from ecosystem-based adaptation

Tim Forsyth

What dilemmas does 'scaling up' community-based adaptation (CBA) bring for risk assessment? For many policy-makers, this is a challenging question. The growing debate about CBA is based on two important, yet not always harmonious, trends in environmental science and policy. The first is the need to address serious risks posed by predicted climate change. The second is the movement to include local communities and citizens in decision-making and risk assessment. At times, these trends appear to work against each other.

At a general level, some analysts have referred to the 'adaptation paradox' as a way to describe the difficulty of addressing climate change as a global concern when vulnerability to climate change is experienced locally (Ayers 2011). But there are additional paradoxes in how 'communities' can perceive or respond to risks that are either outside their historic experiences, or where policies might include actions that seek to change local behaviour and expectations.

Moreover, there is already a trend for climate change policy to emphasise uniform approaches to risk, rather than acknowledge the diversity of local experiences. Official policy documents within the Intergovernmental Panel on Climate Change (IPCC) often base projections of risk upon predicted changes in greenhouse gas concentrations alone. For example, the IPCC distinguishes between 'planned adaptation' and 'autonomous adaptation', which refer to whether adaptation is the result of deliberate policy decisions or 'adaptation that does not constitute a conscious response to climatic stimuli but is triggered by ecological changes in natural systems and by market or welfare changes in human systems' (IPCC 2007: §§5.5.1–5.5.2). Yet both definitions of adaptation make assumptions about how communities experience predicted climate changes as problematic without considering the processes by which 'problems' emerge.

This chapter explores ways to integrate CBA with environmental risk assessment. In order to do this, the ability to define (or redefine) 'risk' needs to become an integral part of the political process of implementing climate change policy and adaptation at the local level. 'Scaling up' CBA does not mean local experiences of risk and local responses should be replicated from place to place. Rather, upscaling CBA implies building the local capacity to identify both risks and solutions flexibility, and hence incorporate the perspectives and needs of vulnerable people within strategies to adapt to increasing greenhouse gas concentrations.

In order to achieve this objective, this chapter argues that environmental policy needs to adopt insights from deliberative politics, which are increasingly being applied to ecosystem management and environmental science. The key objective of inclusive risk assessment is to understand the dynamic and contextual nature of risks under climate change, and to ensure that climate change policies do not exclude potential outcomes by predefining both the nature of risks and solutions. The chapter uses examples of scaling up relating to ecosystem-based adaptation to climate change. But first it reviews some general problems relating to risk assessment and adaptation to climate change at different scales.

Risk assessment and adaptation to climate change

Discussions about risk assessment for adaptation to climate change often focus on a tension between two key approaches in how to define climate risk, and which implicitly refer to different scales. Burton (2009: 89) labelled these approaches as the 'pollutionist' and 'development' perspectives on adaptation. The pollutionist approach sees climate change as essentially a systemic problem, where increasing global concentrations of greenhouse gases will create a number of predictable hazardous events such as changes in precipitation, floods, storms, rising sea levels (and associated salinisation), etc. The 'pollutionist' perspective, therefore, 'is important in deciding what level of greenhouse gas pollution can be tolerated in the atmosphere'. Adaptation under the pollutionist perspective tends to be focused on what are considered the impacts of marginal (or additional) increases in greenhouse gases.

In contrast, however, the 'development' perspective on adaption aims 'to incorporate adaptation to climate in development planning and implementation'. Rather than being systemic for the globe, this perspective focuses on the local conditions that make people or locations vulnerable to increased greenhouse gas concentrations. Consequently, climate 'risk' under the development perspective is defined more in terms of societies' abilities to live with incremental changes in greenhouse gases, and therefore relates more to social vulnerability to climate events (Schipper 2007; Lemos *et al.* 2007).

Much discussion about the risks posed by climate change is therefore divided between those analysts (usually physical scientists) who emphasise the increase in hazardous events, and those (such as specialists in international development) who discuss how impacts can be reduced by protective measures, or by changing activities to those unaffected by events (Ayers and Dodman 2010). Generally speaking, the pollutionist approach was adopted more commonly in earlier discussions of climate change. For example, the IPCC *Second Assessment Report* defined adaptation in terms of systemic physical science: 'the degree to which adjustments are possible in practices, processes, or structures of systems to projected or actual changes of climate' (Watson *et al.* 1996: 16). An early

Dictionary of Global Climate Change (Maunder 1992) did not refer to vulnerability at all.

Yet the pollutionist perspective is still prominent in discussions of how to adapt to climate change. In particular, the pollutionist perspective defines the cause of risk as additional units of greenhouse gas concentrations, and consequently the response to risk is defined in terms of the 'impacts' directly attributable to those concentrations. For example, one impacts analyst wrote: 'Measuring vulnerability can only be done when impacts are known ... measuring the impact of climate change at a country level depends on the downscaling and calibration of global climate models (Tellam 2007: 16).

Nobel-Prize winner Elinor Ostrom has interpreted 'adaptation' as the changes necessary to reduce additional greenhouse gas concentrations, rather than reducing the impacts of those concentrations. In a briefing paper for the World Bank's *World Development Report* on the subject of 'coping with climate change' (Ostrom 2009: 1), she only frames climate change policy in terms of reducing emissions (or mitigation) alone. She writes: '[the chief challenge for policy is] the potential for building a more effective way of reducing green house gas emissions at multiple levels'.

But specialists in social vulnerability and development argue that vulnerability can be measured before any climate impacts by assessing social resilience. Social resilience can be assessed through the diversity of livelihoods that might be affected by climate change; the ability of societies to adapt; and the range of options available to people to make a living. Moreover, social analysts argue that the pollutionist approach, with its focus on managing additional greenhouse gas concentrations, overlooks other appropriate adaptation strategies. For example, Lemos and Boyd (2010: 97) argued: 'through additionality, adaptation policy at the global level divides and circumscribes processes that are indivisible at the local level and, in practice, disables the opportunities for complementaries and synergies in adapting to climate change'.

Indeed, some of these points were made in the early 1990s when the Indian think tank the Centre for Science and Environment argued that global climate change policy should not be based on national levels of greenhouse gas emissions (Agarwal and Narain 1991). These authors argued that national levels hid the fact that per capita carbon usage varied greatly between poorer and richer countries, and that current measurements of emissions hid the fact that richer countries had already undertaken industrialisation or deforestation some decades or centuries before poorer countries. Focusing only on additional greenhouse gas emissions, therefore, did not indicate their origin, or how historic emissions in developed countries has helped them become more resilient to their impacts.

Some analysts now acknowledge these complexities. In recent years the concept of 'adaptation science' has emerged as a form of analysis to identify how adaptation might occur. One early paper defined adaptation science as three questions: 'Adaptation to what? Who adapts? And how does adaptation occur?' (Smit *et al.* 1999). These questions indicate that adaptation need not only be to the direct impacts of increased greenhouse gas concentrations, but also involves social questions of how, and by whom. In this respect, adaptation science is part of a growing field of so-called 'problem-oriented sciences' such as sustainability science, which aim to achieve successful outcomes to urgent socially defined problems rather than applying pre-existing disciplines alone (Clark 2007).

Yet, there are still important challenges for identifying 'problems' within scientific research. By definition, a socially defined problem implies a normative judgement about which outcome is acceptable or not, but classic approaches to 'science' try to avoid normativity. Similarly, problems are not fixed in time, but evolve and change. They are also rooted in local conceptions of what is appropriate and can conflict with wider norms about global environment. Consequently, much 'local' risk assessment can be influenced (knowingly or unknowingly) by ideas that are adopted globally – or ideas represented as 'global' but which only have local roots (Jasanoff and Long 2005).

Can CBA upscale risk assessment?

CBA offers a number of potential solutions to these problems of scale in risk assessment, as well as some challenges. CBA is part of a wider trend in environmental policy towards community-based natural resource management (CBNRM), and accordingly provides more ability for communities to manage resources. It also reflects a trend towards participatory development, which focuses more on local insights and interpretations of environmental change and how this presents problems.

For example, much research on CBNRM in recent years has highlighted how local adaptive practices can exist in the face of population growth or resource scarcity (Batterbury and Forsyth 1999). These studies have often shown that predictions of environmental crisis or collapse can be overstated because local communities adopt practices that maintain the fertility of soil, avoid erosion, or manipulate forest cover to maximise livelihood options, even if outsiders have predicted environmental crisis. Some of these activities include indigenous practices such as the Wolla people's use of soil mounds in Papua New Guinea to increase soil fertility (Sillitoe 1998). They can also include the realisation that some apparent causes of land degradation need not be damaging if local people use them appropriately. For example, some hill farmers in the Nepal Middle Hills use landslides fortuitously in order to facilitate the construction of rice terraces and to turn over old topsoil (Kienholz *et al.* 1984).

In turn, these insights have informed policy approaches aiming to enhance local livelihoods. Sustainable Livelihoods Approaches, for example, have urged that 'poor people themselves must be key actors in identifying and addressing livelihood priorities. Outsiders need processes that enable them to listen and respond to the poor' (Ashley and Carney 1999: 7). 'Sustainable livelihood approaches call for constant questioning of common assumptions and repeated reference to the effects of policy and actions on the livelihoods of the poor' (Carney 2003: 32). In more analytical language, these statements mean there is a need to distinguish between the flow of ecosystem services and the underlying stock of natural capital. Services are, by definition, outputs from ecosystems that are socially defined and useful for different purposes. Natural capital is the underlying land and other biophysical properties. Flows of ecosystem services may be imperfectly related to stocks of natural capital, and there is a need to separate the two (Vira and Adams 2009).

CBA, in principle, is a means of increasing local input into defining 'how' and for 'what' adaptation occurs (following the questions of Smit *et al.* 1999). Yet there are some obvious limitations for CBA for global climate change. Much research about CBNRM has focused on how communities have managed resources that are essential to their livelihoods in previous generations. CBA, however, is not simply adapting to existing climate variability, but involves anticipating changes that are not yet experienced locally. Consequently, most discussions of CBA under the United Nations Framework Convention on Climate Change (UNFCCC) focus only on adaptations to predicted risks under IPCC projections, rather than adaptation to existing climate variability. Moreover, migration away from land is considered an adaptation failure under UNFCCC discussions, even if historically migration has been considered a form of adaptation to the exhaustion of resources (Head 2010).

The UNFCCC also defines other forms of local adaptation within the terms of immediate impacts arising from greenhouse gas concentrations. Autonomous adaptation, for example, is defined as 'adaptation that does not constitute a conscious response to climatic stimuli but is triggered by ecological changes in natural systems and by market or welfare changes in human systems'.[1] It can include practices such as altering agricultural inputs, introducing water-managing technologies, altering cropping cycles or diversifying economic activities. According to this definition, autonomous adaptations are 'largely extensions or intensifications of existing risk-management or production-enhancement activities', which can 'have substantial potential to offset negative climate change impacts and take advantage of positive ones'.

Critics have argued that this definition of autonomous adaptation overlooks pre-existing social institutions and practices, as well as access to technologies and markets, which influence how vulnerable groups respond to new risks (Forsyth and Evans, 2013). Moreover, adaptive responses at the local level are not always driven by environmental changes per se, but by how these changes present hazards for vulnerable people's livelihoods and assets (Carney 2003: 32; Kuruppu 2009; Ribot 2010). Accordingly, various analysts have argued that adaptation among vulnerable populations 'should be done with a deeper awareness of the social, economic, cultural, and political factors that frame their actions, incentives, opportunities, and limitations for action' (Christoplos *et al.* 2009: 3), and that 'adaptation always has, and arguably should, refer to more than just responses to climate change' (Sabates-Wheeler *et al.* 2008: 53).

CBA, therefore, has the potential to offer important new insights into the three central questions of adaptation science: Adaptation to what? Who adapts? And how is adaptation achieved? (Smit *et al.* 1999). Yet some official approaches to adaptation limit some of these questions by assuming that risks come only from the immediate impacts of greenhouse gas concentrations (such as floods and storms) rather than in local vulnerability, or the meaning of these events for affected people. Yet CBA offers the opportunity to integrate different scales of risk assessment by considering how global changes in greenhouse gas concentrations are experienced locally, in ways that are identified locally.

This chapter now considers these challenges in the emerging field of ecosystem-based adaptation, and the challenges of using community-based approaches.

Ecosystem-based approaches to adaptation

Ecosystem-based approaches to adaptation (EBA) integrate the use of biodiversity and ecosystem services into an overall strategy for adaptation to climate change (Munroe *et al.* 2011; Waters 2010). Typical examples of EBA include using vegetation to provide coastal defence; sustainable management of wetlands and floodplains; using forests and land cover to regulate water flows; and establishing agroforestry systems to provide livelihoods and forest cover. Some organisations overtly promote EBA (e.g. the United Nations Environment Programme, UNEP), but there are also many de facto EBA schemes.

EBA offers a number of advantages. First, it can protect resources that are essential for livelihoods and for protecting other resources and assets threatened by events such as floods and storms. Second, these actions can replace or complement other adaptation actions such as building sea walls or storm ducts, in ways that are more consistent with natural ecosystems. Third, it can complement many existing environmental initiatives such as protecting forests. Fourth, restoring ecosystems can have additional benefits through carbon sequestration. Fifth, enhancing ecosystems can be much more affordable than expensive investments in infrastructure (UNEP 2010: 1). In principle, EBA offers opportunities for integrating various perspectives into adaptation to climate change that involve benefits at the local scale; plus contributions to wider ecosystem services that benefit broader spatial scales.

But there are also some tensions in EBA. Some approaches to risk assessment within EBA have reflected the pollutionist, or systemic, means of identifying problems based on atmospheric greenhouse gas concentrations rather than local experiences of risk. These definitions of problems in turn provide potential sources of conflict with local conceptions of development and livelihoods.

For example, the World Bank (2009) published a report entitled, *Convenient Solutions to an Inconvenient Truth: Ecosystem-based Approaches to Climate Change*, which states: 'Ecosystem-based adaptation complements other climate change responses.... Natural ecosystems are resistant and resilient and provide a full range of goods and ecosystem services ... on which human livelihoods depend' (World Bank 2009: 47).

Yet the report goes on to state: 'Agriculture is already one of the greatest threats to natural ecosystems worldwide' (63). And then, in summarising: 'The world's poorest people, who depend directly on the services that various ecosystems provide, are also the most vulnerable to the effects of climate change' (78).

Clearly, these statements represent a vision of risk that is based on changes to ecosystems per se, rather than what those changes mean to affected people. Representing 'agriculture' as a 'threat to natural ecosystems' seems to miss the point that agriculture can also be a source of livelihoods and resilience – and that many developed nations gained significantly from agriculture (and by transforming their ecosystems at the same time). Similarly, linking 'the world's poorest people' with a dependency on ecosystem services could equally be an argument for addressing poverty, rather than restoring ecosystem services.

The World Bank report then makes specific policy recommendations concerning restoration and maintenance of ecosystems through investments in vegetative land cover. It expresses concern that the Clean Development Mechanism (CDM) does not yet maximise investment in forests and reforestation (80).

Yet this statement also overlooks the long-standing controversy about the use of 'sink' (or forest and land-use-based) projects within the CDM. A similar concern is about the use of reforestation under the CDM or Reduced Emissions from Deforestation and Forest Degradation (REDD+) for industrial tree plantations. Critics argue that large-scale plantation forestry does little to restore biodiversity, and can have knock-on hydrological effects such as reducing water supply, and even enhancing soil erosion (Calder 1999; Bruijnzeel 2004). Forest plantations also do not increase access of land for smallholder agriculture, and hence might restrict livelihood options.

Not all EBA projects make the same statements as in this World Bank report. Indeed, the International Union for Conservation of Nature (Colls *et al.* 2009: 1) reported that EBA 'can be readily integrated into community-based adaptation and addresses many of the concerns and priorities identified by the most vulnerable countries and people'. And the activist group the Climate Action Network (CAN 2009: 9) writes that EBA can achieve work that is: 'locally focused and community based, taking account of traditional knowledge … and that addresses the needs of the most vulnerable countries, communities, ecosystems and livelihoods'.

But there are clearly challenges for risk assessment within EBA that need to be analysed further in order to understand how to avoid tensions between scales in adaptation to climate change; and to identify the role of CBA within this process.

Scaling up community-led ecosystem-based adaptation

How can communities be involved in risk assessment in order to make EBA more useful for local people? And what lessons can be drawn from locally successful CBA that can be transferred to different locations and scales? These important questions reflect the tensions noted above.

First, from the viewpoint of adaptation to climate change:

- Adaptation is not just adaptation to existing climate variability, but the anticipation of future changes.

- Local communities are frequently vulnerable to climate change (especially in developing countries), but have not yet experienced the kinds of climate change predicted.
- Some changes in local practices, such as land use or other economic activities, might be required in order to achieve adaptation.

Yet from the viewpoint of local development objectives:

- Risk is not just indicated by the increased magnitude and frequency of climatic events, but human exposure to them.
- Adaptation is not simply protecting against the direct impacts of climate change such as storms, but can include changing livelihoods and development trajectories that make these direct impacts less relevant.
- Protecting ecosystem services does not simply mean protecting existing land cover, but asking which services are needed, and which actors see them as 'services'.

Two approaches might overcome these tensions.

Diversifying the science of ecosystem services

The first method involves considering how, and with whose influence, ecosystem services are defined. This question relates to spatial scale of services, but also the global reach of normative values about services and successful policy outcomes. It also refers to the applicability of different scientific models of explaining ecosystem services across different contexts and scales.

For example, there are some applications of EBA where all parties agree ecosystem services are valuable. For example, coastal vegetation can protect beaches and agricultural land from surges and erosion. But controversies arise when assumptions are made on behalf of affected people, or when scientific generalisations are used out of context. For example, much research referred to above about environmental crises in developing countries has shown examples of 'received wisdom' about the causes and effects of environmental changes that do not necessarily reflect the local experience of poorer people.

For example, the US government developed the Universal Soil Loss Equation (USLE) after the serious problems of soil erosion experienced in the lower plains of the United States during the so-called 'Dust Bowl' of the 1930s. This equation was designed to predict the causes and effects of erosion for all circumstances based on factors such as slope length, soil erodibility, rainfall and soil conservation measures. Yet, despite the name, this equation has been shown to be anything but universal when applied to developing countries – either because the proportion given to the different parts of the equation do not reflect rainfall intensity, soil formation rates or land management in the tropics, or because 'erosion' might be less of a problem than other causes of declining soil fertility or agricultural failure (Forsyth and Walker 2008).

Similarly, much EBA discusses reforestation as a means of restoring ecosystems and their services. But there is much criticism of monoculture plantations for both biophysical reasons (such as impacts on hydrology and biodiversity), and for development outcomes (by restricting access to agricultural land) (Calder 1999). Much research on the impacts of deforestation on ecosystem services also indicates that the impacts of 'deforestation' on hydrology can simplify various cause-and-effect statements. For example, Bruijnzeel (2004) argues that much hydrological debate assumes that deforestation will decrease the supply of lowland water by leading to soil compaction, which causes flash floods during rainy seasons, and because trees encourage the slow release of water during dry seasons because they act as a store of water. Yet, Bruijnzeel argues that these simple cause-and-effect statements about the impact of deforestation can be achieved via multiple means, and not simply through the removal of trees. For example, landscapes can still provide water supply if the style of agriculture adopted can maintain high levels of soil infiltration – and if governments and local people refrain from building roads that act as fast conduits of water. Under the right conditions, therefore, the removal of trees need not lead to declining water supplies, and landscapes can still provide space for agriculture.

The point of these kinds of examples is that many statements about ecosystem services are like the 'pollutionist' approaches to risk described above. They are 'black-box' summaries of cause and effect that hide various alternative explanations, and which legitimise 'solutions' that might close down other land uses that might assist in local development. The way to avoid these black-box statements is to avoid transferring generalisations about ecosystem services across contexts without analysing whose knowledge and values have helped create them.

Community-based assessments can provide useful insights concerning which ecosystem services are valued, and which proposed solutions might provide advantages or disadvantages for local development. This kind of analysis is similar to participatory land-use planning, which actively involves local people in identifying which ecosystem services are important, and which land might be used for projects such as reforestation with least conflict (Pinkaew 2001: 156). The purpose of this analysis is not to present a zero-sum choice between 'local' or 'outsider' visions of ecosystem services, but to seek to make proposed forms of adaptation more compatible with local needs and experiences. Using diverse knowledge sources, and looking for synergies or contradictions, reduces the reliance on knowledge generated from one source alone, and allows each to be viewed in its own context. This approach is not to champion 'local' knowledge but to seek various solutions to ecosystem services and various 'co-benefits' from each proposed strategy. Indeed, some approaches to EBA already emphasise co-benefits as a way to establish win–win outcomes from adaptation policies (Colls *et al.* 2009).

Diversifying environmental governance

There is also a need to reform adaptation policy processes to allow greater flexibility in order to answer the questions: Adaptation to what? Who adapts? And how is adaptation achieved? Again, these involve resolving tensions between pollutionist- and development-oriented approaches to adaptation; and the implications for implementing adaptation at a local scale.

Table 6.1 shows a simplified summary of three main policy processes currently discussed in environmental research. The column on the left shows the common approach adopted within climate change negotiations based upon the pollutionist model, such as discussed by Ostrom (2009). This style of policy-making identifies a key 'problem' and then seeks to implement the solution at interlocking spatial scales (so-called upscaling and downscaling, or nested institutions). This approach, for example, can be used to implement greenhouse gas mitigation at international, national and sub-national scales, such as proposed under REDD+.

The column on the right summarises the insights from CBNRM, which emphasises local perceptions and development needs. This approach uses terms such as legal pluralism to refer to the interlocking diverse policy frameworks that might come from various formal and informal sources. CBNRM, however, as noted above, might have tensions when connecting with 'global' environmental concerns such as climate change mitigation, or when local people might not yet experience the kinds of risks posed by future climate change.

The column in the middle, however, is an attempt to build a more deliberative approach to risk assessment, where policies are not based on either a 'global' vision of risk, nor an overtly localist approach. Instead, this middle ground allows for some deliberation about the nature of risk and the search for inclusive processes about which risks, and which solutions, are considered. The chief problem with this approach is to achieve an inclusive public debate where affected people might be relatively voiceless compared with large organisations or governments. But a form of EBA based on deliberation might be more useful for adaptation, and for local development objectives, than policy based upon predefined normative assumptions about what landscapes or interventions are useful for adaptation.

Conclusion

For many observers, CBA to climate change presents an 'adaptation paradox' because the nature of risk is global (or driven by atmospheric greenhouse gas concentrations), but the agency for adaptation is undertaken by local actors (Ayers 2011). Consequently, there are fears that CBA is ultimately limited, or cannot be 'scaled up' to wider spatial scales, or transferred from one location to another.

This chapter has argued that these concerns are misplaced. In order to scale-up CBA and apply its lessons more generally, there is a need to see CBA as a

Table 6.1 Different approaches to multi-level governance

	Policy objectives driven more from above	←→	*Policy objectives driven more from the local level*
Type of governance	Nested institutions	Deliberative networks and institutions	Legal pluralism, including CBNRM
Reference	Ostrom 2009	Rodela, 2012	Schroeder 1999
Main mechanisms	Actors create rules for enforcement and monitoring, in coordination with higher authorities ('Common Property Regime')	Policy is shaped by open discussion and participation by various stakeholders, aiming to empower localities to take responsibility for policy	Recognise co-existence of formal and informal governance regimes at different scales, as practised by communities
Main advantage	Clearly defined rules	Dynamic, localised and learning	Reflects the complexity of local rule-making
Main disadvantage	Does not always acknowledge local perceptions of forests or local political processes	Civil society might be dominated by elites and the state.	Does not always relate to urgent 'global' tasks such as controlling emissions
Implications for equity, efficiency and effectiveness	Can be effective and efficient where rules are agreed; but all three Es can be damaged where local practices are ignored	High potential for three Es, but if process is slow and conflictual, efficiency is low	Equity might be high but low effectiveness?

Source: adapted from Forsyth 2009.

space for the local identification and response to risk; rather than a fixed response to a fixed risk. In other words, CBA offers a platform for vulnerable people to participate in discussing how, and for whom, hazardous events linked to climate change actually become risks for specific people and places, and then to identify appropriate solutions for these risks. These solutions need not be based only upon responses to the immediate impacts of these hazards (such as by building sea walls to reduce impacts of storm surges), but also considering how current livelihoods make certain events hazardous, and how different configurations of livelihoods would in effect change the nature of hazards (Forsyth and Evans, 2013).

CBA therefore provides a potential arena for local deliberation about risks and solutions. But this is not all that is required. In order for CBA to be an effective cross-scale policy response, there is also a need to reconsider the means by which global risks are defined, and therefore might predefine what adaptation is for. For example, is it appropriate that the World Bank (2009) produces recommendations for adaptation to climate change that implies that agriculture is part of the problem? Or that restoring 'natural' ecosystems is a way to reduce vulnerability of poorer people? This chapter argues that these kinds of statements put the cart before the horse; they misplace the origins of risk and the potential negative impacts of so-called adaptation strategies. Yet the potential for CBA and EBA to enhance local development and protect against climate change is there.

Accordingly, there needs to be more attention to some of the obvious tensions in risk assessment under current approaches to CBA.

- CBA is still too frequently framed within the so-called 'pollutionist' approach to adaptation, which measures risk in terms of incremental units of global greenhouse gases. Instead, there needs to be a greater analysis of how development can reduce vulnerability.
- Ecosystem services are frequently discussed in ways that equate flows of services with stocks of natural capital. Instead, there is a need to distinguish the two, and to see many beliefs about cause and effect regarding ecosystem services be based on selective experiences, empiricism and land uses that do not always travel to poorer societies.
- Community-based assessment is still too frequently considered to be an alternative to, rather than a complement to, global assessments of climate risks. Involving local people in risk assessment might not reveal knowledge of future climate changes. But participatory methods can help ensure that policies are not unnecessarily restrictive of livelihoods, or based on assumptions about ecosystem services that are out of context.

A more deliberative approach to risk assessment – where different perspectives on risks and services are considered inclusively rather than exclusively – is a way to ensure that multiple risks (at multiple scales) are addressed by CBA. Integrating local concerns into climate change risk assessment does not just mean reducing risk to what local people believe is happening to climate. It means making spaces to allow the assessment of risks from various perspectives in ways that consider global drivers of risk alongside local causes of vulnerability and ensuring that adaptation can reduce local vulnerability rather than make it worse.

Note

1 www.ipcc.ch/publications_and_data/ar4/wg2/en/annexessglossary-a-d.html.

References

Agarwal, A. and Narain, S. (1991) *Global warming in an unequal world*, New Delhi: Center for Science and Environment.

Ashley, C. and Carney, D. (1999) *Sustainable livelihoods: Lessons from early experience*. London: Department for International Development.

Ayers, J. (2011) 'The adaptation paradox', *Global Environmental Politics* 11 (1): 62–88.

Ayers, J. and Dodman, D. (2010) 'Climate change adaptation and development I: the state of the debate', *Progress in Development Studies* 10 (2): 161–168.

Batterbury, S. and Forsyth, T. (1999) 'Fighting back: human adaptations in marginal environments', *Environment: Science and Policy for Sustainable Development* 41 (6): 6–9.

Bruijnzeel, L.A. (2004) 'Hydrological functions of tropical forests: not seeing the soil for the trees?', *Agriculture, Ecosystems and Environment* 104 (1): 185–228.

Burton, I. (2009) 'Climate change and the adaptation deficit,' pp. 89–98 in Schipper, E. and Burton, I. (eds) *The Earthscan reader on adaptation to climate change*, London: Earthscan.

Calder, I. (1999) *The blue revolution: land use and integrated resource management*. London: Earthscan.

CAN (2009) 'Briefing paper on the role of ecosystems in adaptation', May 2009, Climate Action Group.

Carney, D. (2003) *Sustainable livelihoods approaches: progress and possibilities for change*, London: DFID.

Christoplos, I., Anderson, S., Arnold, M., Galaz, V., Hedger, M., Klein, R.J.T. and Le Goulven, K. (2009) *The human dimension of climate adaptation: the importance of local and institutional issues*, Stockholm: Commission on Climate Change and Development.

Clark, W. (2007) 'Sustainability science: a room of its own', *Proceedings of the National Academy of Science* 104 (6): 1737–1738.

Colls, A., Ash, N. and Ikkala, N. (2009) *Ecosystem-based adaptation: a natural response to climate change*. Gland: IUCN.

Forsyth, T. (2009) 'How to participate, integrate and coordinate? REDD and multi-level, multi-actor governance', pp. 113–122 in Angelsen, A. and Brockhaus, M. (eds), *Realising REDD+: national strategy and policy options*. Bogor: CIFOR.

Forsyth, T. and Evans, N. (2013) 'What is autonomous adaptation? Resource scarcity and smallholder agency in Thailand', *World Development* 43: 56–66.

Forsyth, T. and Walker, A. (2008) *Forest guardians, forest destroyers: the politics of environmental knowledge in Northern Thailand*, Seattle: University of Washington Press.

Head, L. (2010) 'Cultural ecology: adaptation – retrofitting a concept?', *Progress in Human Geography* 34 (2): 234–242.

IPCC (2007). *Climate change 2007: Working Group II: impacts, adaptation and vulnerability: Glossary*. Available from http://www.ipcc.ch/pdf/glossary/ar4-wg2.pdf (accessed 19 September 2013).

Jasanoff, S. and Long, M. (eds) (2005) *Earthly politics: local and global in environmental governance*, Cambridge, MA: MIT Press.

Kienholz, H., Schneider, G., Bichsel, M., Grunder, M. and Mool, P. (1984) 'Mapping of mountain hazards and slope stability', *Mountain Research and Development* 4(3): 247–266.

Kuruppu, N. (2009) 'Adapting water resources to climate change in Kiribati: the importance of cultural values and meanings', *Environmental Science and Policy*, 12, 799–809.

Lemos, M. and Boyd, E. (2010) 'The politics of adaptation across scales: the implications of additionality to policy choice and development', pp. 96–110 in Boykoff, M. (ed.) *The politics of climate change: A survey*, London: Routledge.

Lemos, M., Boyd, E., Tompkins, E., Osbahr, H. and Liverman, D. (2007) 'Developing adaptation and adapting development', *Ecology and Society* 12 (2): 26.

Maunder, W. (ed.) (1992) *Dictionary of global climate change*, London: UCL Press.

Munroe, R., Doswald, N., Roe, D., Reid, H., Giuliani, A., Castelli, I. and Moller, I. (2011) *Does EBA work? A review of the evidence on the effectiveness of ecosystem-based approaches to adaptation*. London: IIED.

Ostrom, E. (2009) 'A polycentric approach for coping with climate change', *World Bank Policy Research Working Paper* 5095, Washington, DC: World Bank.

Pinkaew, Laungaramsri (2001) *Redefining nature: Karen ecological knowledge and the challenge to the modern conservation paradigm*. Chiang Mai: Regional Center for Social Science and Sustainable Development, Faculty of Social Science, Chiang Mai University.

Ribot, J.C. (2010) 'Vulnerability does not just come from the sky: framing grounded pro-poor cross-scale climate policy', pp. 47–74 in Mearns, R. and Norton, A. (eds), *Social dimensions of climate change: equity and vulnerability in a warming world*. Washington, DC: World Bank.

Rodela, R. (2012) 'Advancing the deliberative turn in natural resource management: an analysis of discourses on the use of local resources', *Journal of Environmental Management* 96: 26–34.

Sabates-Wheeler, R., Mitchell, T. and Ellis, F. (2008). 'Avoiding repetition: time for CBA to engage with the livelihoods literature?', *IDS Bulletin*, 39 (4): 53–59.

Schipper, E.L.F. (2007) 'Climate change adaptation and development: exploring the linkages', *Tyndall Centre for Climate Change Research Working Paper 107*.

Schroeder, R. (1999) *Shady practices: agroforestry and gender politics in The Gambia*, Berkeley, CA: University of California Press.

Sillitoe, P. (1998) 'It's all in the mound: fertility management under stationary shifting cultivation in the Papua New Guinea highlands', *Mountain Research and Development* 18 (2): 123–134.

Smit, B., Burton, I., Klein, R. and Street, R. (1999) 'The science of adaptation: a framework for assessment', *Mitigation and Adaptation Strategies for Global Change* 4: 199–213.

Tellam, I. (2007) 'Adaptation targets', *Tiempo* 64: 14–16.

UNEP (2010) *Using ecosystems to address climate change: ecosystem based adaptation*, Nairobi: United Nations Environment Programme Regional Seas Programme.

Vira, B. and Adams, W. (2009) 'Ecosystem services and conservation strategy: beware the silver bullet', *Conservation Letters* 2: 158–162.

Waters, J. (2010) 'Ecosystems and adaptation', *Tiempo* 77: 3–7.

Watson, R.T., Zinyowera, M.C. and Moss, R.H. (eds). (1996) *Climate change 1995: impacts, adaptations and mitigation of climate change: scientific-technical analyses.* Cambridge: Cambridge University Press.

World Bank (2009) *Convenient solutions to an inconvenient truth: ecosystem-based approaches to climate change*, Washington, DC: World Bank.

7 Scaling up the use of tools for community-based adaptation*

Issues and challenges

Tine Rossing, Agnes Otzelberger and Pascal Girot[1]

Over the years, donor agencies, international organisations and NGOs have issued an array of guidance documents and tools on climate change adaptation and its integration into development programming, covering a broad spectrum between 'generic mainstreaming guidance' (UNDP 2010: 6), on the one hand, and specific tools for adaptation, on the other. Few of them provide guidance on how to scale-up community-based adaptation (CBA) knowledge and practical lessons. This gap is an important concern, as there is a growing need for CBA to reach a critical mass, both horizontally across a larger number of vulnerable communities and vertically to support policy traction at local, national and international levels.

This chapter discusses the particular challenges of scaling up CBA through tools and methods. For tools targeting CBA[2] these challenges are numerous. A first set of challenges concerns building and sharing knowledge and scalable experiences of CBA. In our view, these processes and steps constitute a substantial precondition for the process of scaling up CBA practice itself, which comes with a second set of challenges. Before discussing these in more detail, however, we take a brief look at efforts to take stock of adaptation tools made to date, and define what we mean by 'scaling up' in the context of this chapter and of CBA tools.

Taking stock of existing adaptation tools

A few recent reviews (e.g. Tanner and Guenther 2007; GTZ 2009; Olhoff and Schaer 2010; UNDP 2010; Hammill and Tanner 2011, Schipper 2009) have taken stock of what has become available on the 'tools market' so far. Existing tools are highly diverse in terms of their characteristics, target audiences, aims and terminologies, but they respond most often to institutional prerogatives rather than being demand driven. The reviews, therefore, highlight a need for harmonisation between tools (Olhoff and Schaer 2010) and make attempts at systematising the tools into different categories, providing orientation for potential users, while identifying their strengths and gaps. Hammill and Tanner (2011), for instance, establish a new, three-fold typology (see Table 7.1) based on the functions of different tools vis-à-vis the climate risk management process,

Table 7.1 Categories of climate risk management and adaptation tools

Type/characteristics	*Notes*	*Examples from the development community*
1 Process guidance tools Tools that guide users through the identification, gathering and analysis of relevant data and information to: • Identify climate risks to development activities (often using Type 2 tools) • Assess and analyse climate risk management strategies • Evaluate option to integrate climate risk management into development activities	Guide users through the entire CRM/adaptation process (e.g. from awareness-raising to monitoring and evaluation), or just one or several steps in the process (e.g. assessing current and future climate risk). Majority are available as documents, some are available as computer programs.	• Adapting to Coastal Climate Change: A Guidebook for Development Planners: www.crc.uri.edu/index.php?actid=366 • BMZ Environment and Climate Assessment: www.gtz.de/climate-check • CEDRA: http://tilz.tearfund.org/Topics/Environmental+Sustainability/CEDRA.htm • CRiSTAL: www.cristaltool.org • ORCHID: www.ids.ac.uk/climatechange/orchid • USAID Guidance Manual: www.usaid.gov/our_work/environment/climate/policies_prog/adaptation.html
2 Data and information provision tools These tools generate or present data and information on: • Primary climate variables and projections (e.g. temperature, rainfall trends) • Secondary climate impacts (e.g. flood maps, crop yields) • Vulnerability and response options (e.g. poverty maps, example adaptation options)	Generally require some computer capacity or internet access. They tend to be databases, modelling programs, mapping and visualisation tools.	• CI-Grasp : www.ci-grasp.org • Climate Wizard: www.climatewizard.org • Climate Change Explorer Tool: www.weadapt.org/wiki/The_Climate_Change_Explorer_Tool • PRECIS: www.precis.metoffice.com • SERVIR: www.servir.net • World Bank CC Knowledge Portal: climate, impact and socio-economic data: http://sdwebx.worldbank.org/climateportal
3 Knowledge-sharing tools Platforms and networks that offer adaptation practitioners a virtual space for information and experiences related to climate risk and adaptation. These spaces allow users to: • House or store information and knowledge • Share it with other interested users • Interact with other users to develop or advance ideas, approaches, tools, monitoring, etc.	Typically knowledge platforms, increasingly reliant on Web 2.0 functionality and user-generated content. Can be important for *validation* of Type 1 and Type 2 tools, as these platforms can offer a space for user feedback and offer some sort of quality control mechanism. Help build a *community of practice* around climate change adaptation.	• Adaptation Learning Mechanism: www.adaptationlearning.net • AfricaAdapt: www.africa-adapt.net • Climate Adaptation Knowledge Exchange: www.cakex.org • Climate One Stop: http://arcserver4.iagt.org/climate1stop • ELDIS Resource Guide on Adaptation: www.eldis.org/go/topics/dossiers/climate-change-adaptation • weADAPT platform: www.weadapt.org • World Bank CC Knowledge Portal: http://sdwebx.worldbank.org/climateportal/

Source: Hammill and Tanner 2011.

dividing tools into (1) data and information provision; (2) knowledge sharing; and (3) process guidance tools. Their review concludes that most tools 'target particular niches' (Hammill and Tanner 2011: 7), are unable to look at multiple stressors (beyond climate change) and miss out on guiding the crucial step between analysis, on the one hand, and designing adaptation interventions, on the other (Hammill and Tanner 2011: 35).[3]

None of these reviews have focused specifically on tools for CBA per se or addressed the challenges related to applying these tools for scaling up CBA, which is the focus of this chapter. CBA is still a relatively recent and rapidly evolving field, and, as such, is undergoing an iterative learning process within the global CBA community, where knowledge about adaptation grows based on practice (Huq 2011). At present, many of its tools are limited in scope or types of risk they take into account, and the scale of their application is often restricted to households or communities, providing only limited guidance on how to engage with broader governance structures and contexts.

Defining scaling up and scaling 'out'

'Scaling up' community-based adaptation can encompass a variety of processes that are both vertical and horizontal in nature. Upscaling, or *vertical* scale-up, first concerns the notion of scaling up from the local or community level to higher levels of decision-making (such as sub-regional or national) within a given country or globally.[4] To date, upscaling climate adaptation in general has mainly been addressed through mainstreaming climate change into sectoral policies and primarily carried out at the national level, as a top-down strategy. One plausible explanation could be that *implementation* of CBA is still gaining ground, as most efforts to date have focused on CBA *planning*. While one challenge for CBA in this top-down process has been and still is the availability of *down-scaled* climate data to inform local-level adaptation strategies (Birkmann *et al.* 2009: 30), 'upscaling' reverses this top-down trend in relying on lessons learned from local change processes to inform decision-making at higher administrative and organisational levels with wider-reaching impact (Larsen *et al.* 2011: v). As such, in theory, CBA tools can facilitate the process of scaling up CBA if they help build experience and document good practices that result in appropriate adaptation decisions at higher levels. Burton also links the scaling up process to a need for enhanced integration of CBA into the wider development processes, highlighting the importance of avoiding stand-alone adaptation initiatives (Burton 2011).

Second, CBA can be expanded over a larger geographical area, which can be defined as '*scaling out*'. Such a *horizontal* scale-up process could involve a small-scale project intervention or initiative, the scope of which is broadened into a larger-scale endeavour (Snapp and Heong 2003). The expansion could involve a larger number of new but replicated community-based initiatives based on the initial intervention. It could also involve a larger number of beneficiaries or a whole region as opposed to one or a few targeted communities.[5]

Tools aimed at supporting such processes, therefore, need to allow for replication of CBA across different contexts or with larger outreach (horizontal and/ or vertical), while remaining both context-appropriate and coordinated with wider policy, legislative, planning and budgeting frameworks needed to sustain such outcomes at scale and over time.[6] As Brooks *et al.* (2011: 7) point out, such tools will also need to consider:

> The long timescales over which many aspects of climate change (and hence much adaptation) will unfold, and the problems of how to evaluate the impacts of adaptation interventions on development outcomes in the face of rapidly evolving stresses and risks that change the contexts in which development takes place (the 'shifting baseline' problem).

Challenges to creating and sharing knowledge for scalable CBA experiences

Successful CBA approaches introduce sustainable development practices that make communities more resilient, both to immediate climate variability and long-term climate change (Huq and Reid 2007; Brooks *et al.* 2011). Their respective measures of success, however, will vary significantly given the highly contextual nature of CBA, as climate impacts ultimately are manifested locally as risk-building processes (Lavell 2004). These occur when climate hazards interact with exposed livelihood assets and differentiated vulnerabilities, resulting in increased climate-related risks.

As such, the very nature of CBA poses a significant challenge to building replicable, scalable case studies. As summed up by Larsen *et al.* (2011: 2) in a recent study: How can lessons be learned from diverse CBA initiatives and shared across different regions, contexts and local realities, to scale up and out CBA experiences? How can highly contextual local change processes inform generic national or sub-national policies and processes which can be implemented in all localities? Also, what are the prospects for a dialogue between policy-makers at the national level who are primarily concerned with drawing generalising conclusions based on local lessons, and local professionals who examine community-based climate adaptation in context? A set of challenges related to building and sharing knowledge and scalable experiences of CBA must be overcome in order to successfully bridge different levels and actors. Good documentation of scalable practices constitutes a precondition and point of departure for the process of scaling up and out CBA practice itself.

Based on experience from other sectors (Larsen *et al.* 2011: 2), such as disaster risk reduction and natural resource management, particularly integrated watershed management, governments tend to upscale lessons from local change processes and integrate them into sectoral policies based on case study research. To foster increased upscaling and outscaling efforts of CBA, it is therefore imperative to develop and apply tools and methods that support the documentation of strong CBA case studies. This said, documenting good practice in an ever-changing context is a difficult task to undertake.

To facilitate the uptake of these case studies, tools and methods should also stimulate dialogue between practice and policy, i.e. the local climate adaptation initiatives and the institutional environment these are embedded in. This is equally the case for outscaling, as to broaden the geographical area of a given adaptation intervention will equally require such linkages. Yet, outscaling can also be promoted through the agency of multi-stakeholder platforms, such as producer organisations and other practitioner networks. In Central America, for example, ACICAFOC, a regional agro-forestry producers federation, scaled out the spread of rain water harvesting techniques through local producer organisations.[7]

Lack of tools to gather the climate information for CBA planning, design and feasibility assessments

Information on potential impacts of climate change must be location-specific and must be provided to communities in an appropriate format (Huq 2011). The use of meteorological and climate forecast services for short-, medium- and long-term scales enable farmers to move beyond short-term risk response strategies towards more forward-looking adaptation. Access to reliable local information about climate projections and seasonal forecasts is, therefore, a fundamental requirement for CBA planning and implementation. Yet, at present, the tools and methods required to access and process such information are only in the initial stages of development (Percy, pers. comm.; Ewbank, pers. comm.).

In least developed countries (LDCs) in particular, the availability and quality of this crucial information in meteorological departments and services is poor and only slowly increasing. In addition, these departments have not been part of the development arena and are not necessarily resourced to provide information in a format and at a cost which makes it accessible for communities and local-level service providers (such as agriculture extension workers and providers of livestock health, credit and insurance) for decision-making concerning local adaptation planning and implementation (Percy, pers. comm.; Ewbank, pers. comm.). Equally, from a community perspective, land-user communities often have localised knowledge concerning past climate trends and seasonal forecasts, which are valuable contributions to add to meteorological information for local decision-making. Local actors can also contribute to improved weather data by keeping rainfall and temperature records in a local system (Percy, pers. comm.).

Mechanisms for improved integration of meteorological services into development and adaptation planning processes, and for systems of two-way communication between meteorological services and local community actors, have started to emerge (Glantz 1988, 1990, 2003; Ewbank, pers. comm.). Also, the 'Adaptation Learning Programme for Africa' (Percy, pers. comm.), is currently developing mechanisms that will enable meteorological services to identify what climate-related information is needed, when and in what format, at the local level. This includes support to new participatory scenario planning methods, including systems for two-way communication between the meteorological stations and local community actors. It is recognised that robust institutions developing and implementing these systems will be essential for sustainability and for ensuring the system is useful for the stakeholders most vulnerable to negative climate impacts. Other concerns related to consistency and quality control also explain why many services have been reluctant to include locally gathered climate data (Glantz 2003).

Another set of fundamental tools and methods that currently need further development are the ones needed to assess the feasibility of proposed adaptation actions (McCarthy *et al.* 2001; Berkes and Jolly 2001). Concerning CBA, a key reason for this gap is that while CBA planning has made great strides, implementation remains in its incipient stages. What is also needed is a tool that can measure change in adaptive capacity. As stated by Percy (pers. comm.), adaptation to ongoing climate change requires adaptive capacity 'which means being able to plan and act in response to experience, information and anticipation of changing and uncertain circumstances'.

Incompatibility and insufficient links between different types of tools

At present, different types of adaptation tools are not sufficiently linked with each other. Hammill and Tanner (2011) point out that, in an ideal world, distinct tools addressing different aspects of adaptation to climate change would be compatible with each other, thereby providing complementary guidance on how to plan, implement, document, upscale and outscale CBA initiatives. In reality, the tools are not sufficiently interlinked to ensure a continuous process. Outputs from individual tools at one stage in the project cycle do not easily lend themselves to being processed through a different tool at the next stage. A particular gap exists between climate information, vulnerability analysis and community adaptation planning – practitioners struggle to make the steps between understanding climate risks and adequately planning actions to address them. Hammill and Tanner (2011) find that process guidance tool users, for example, rarely consult the outputs of information provision tools. They attribute this disconnect to the diverging profiles, capacities and needs of the users for each category (Hammill and Tanner 2011: 37). Also, different definitions for and uses of key climate terminology like 'risk' and 'screening' are currently contributing to confusion on how to identify and apply appropriate tools for upscaling adaptation (Hammill and Tanner 2011: 38).

Additionally, an assessment of tools for community-based risk management, including CBA, carried out by Action Against Hunger, revealed various reasons for fragmentation – for example, different institutional settings and set-ups, or diverging visions of risk (Mitchell and Otzelberger 2010). Such fragmentation makes it difficult to combine these tools in support of up- and outscaling of CBA initiatives. The assessment also highlighted that even CBA-specific tools are often aimed at highly specific target groups. FAO's e-learning tool *Planning for*

Community-Based Adaptation to Climate Change, for example, targets extension workers (Ricoy 2011).

Efforts to guide linkages and fill gaps run a challenging race against the fast-paced environment of climate change adaptation: the development of CARE's digital CBA toolkit,[8] which seeks to connect different tools from both CARE and other organisations around the CBA project, was faced with a rapidly evolving landscape of CBA tools and experiences. By the time the toolkit was ready for release, there were new tools available that could have been integrated and linked. The process of identifying the right tool, however, is an important learning experience in its own right. As lessons are gained from field testing CBA tools, there should always be openness to expand upon and improve existing guidance.

Harmonising efforts and pooling resources to develop a 'one-size-fits-all' tool or set of tools for upscaling and outscaling does not appear to be the right way forward, given the importance of developing processes tailored to the specific needs and contexts of a given CBA initiative. For example, while up- and outscaling is not necessarily done separately, the two processes are not identical. They may overlap to some degree, but each process involves its own set of stakeholders and challenges that need to be addressed by different tools and approaches. As Ribot (2009: 18) suggests: 'Principles to govern climate action must be designed around the processes that shape vulnerability and the actors and organizations with authority and power to make decisions that can change these processes.' In this sense, improved coordination and collaboration should not be limited to tool users and developers, but enable 'diverse voices to feed into the climate risk management process' which would 'enhance ownership and promote development of more appropriate and effective adaptation options' (Hammill and Tanner 2011: 37).

Capacity constraints and knowledge-sharing challenges

Insufficient capacity to apply tools constitutes another particularly significant challenge to creating and documenting CBA experiences. Without proper guidance, inappropriate application of adaptation tools may inadvertently contribute to maladaptation (Olhoff and Schaer 2010). Hammill and Tanner (2011: 26) note that tool *developers* have experienced a growing demand for training, and tool *users* maintain that such training was critical to their experience in applying the tools.

At the community level, where climate risk may be a significant, but far from the only, issue to be addressed, practitioners have to juggle multiple and complex demands. The novel and complex nature of CBA makes it difficult for its tools to be used right away by these practitioners without any prior induction or training. Yet, even when training in tools application is provided, experience shows that such training is often rushed and delivered without the proper background, and, as a result, it lacks the desired effects. There is, in this sense, a trade-off between ambition, speed and resources. Tools users need to understand the theory behind the design of these tools, so that the learning becomes an iterative and transformative process. There is, therefore, a need for real investments in building capacity to support well-informed use and continued innovation through documented experiences. This need, however, often cannot be met due to lack of funding and time (Gambarelli 2011).

While such capacity development is key, it is also important to take on the underlying causes of capacity constraints. Once trained, local field-based technicians are often lured by more lucrative positions in the cities, as CBA skills are in increasing demand. As a result, there is a significant brain drain among practitioners with experience in how to apply CBA tools at the field level, putting limits on the speed and level at which additional and larger initiatives can be planned by the organisations in question. Training, therefore, needs to be accompanied by strategies to retain and motivate local technicians and empower local organisations to provide their staff with adequate work conditions and means to apply their knowledge in the most effective manner. This can be achieved through the building of platforms and networks that offer adaptation practitioners a virtual space for information and experiences related to climate risk and adaptation. These spaces allow users to (1) house or store information and knowledge; (2) share it with other interested users; and (3) interact with other users to develop or advance ideas, approaches, tools, monitoring, etc. (Hammill and Tanner 2011: 14). In addition, knowledge-sharing tools can help connect different, but complementary tools, addressing the challenges raised above. Through increased sharing and exchange, such symbiotic linking can be further enhanced.

The widespread proliferation of knowledge-sharing tools, such as climate information portals and knowledge platforms, can also be counterproductive. As many portals continue to be designed and operated in isolation, there is an increasing risk of duplication of effort and reinvention of wheels or 'silo thinking' (Barnard 2011: 2). One way to address this challenge would be to establish an online central clearinghouse for adaptation tools, where users would gain rapid exposure to the range of different tools 'out there', which, in turn, would allow users to better identify which tool or combination of tools match their particular needs (Hammill and Tanner 2011: 41). Such a clearinghouse mechanism could also support and encourage a broader knowledge- and experience-sharing process among adaptation to help inform, support and refine the planning and implementation of the different steps or approaches related to outscaling and upscaling. To date one of the closest attempts of such collaboration is the approach that is pioneered by WeAdapt, who makes its web platform available to partners to add their content, sign up their users and put their own logo and branding across the top. Not only is this approach cost-effective, it also creates scope for information-sharing between different communities that share the same platform (Barnard 2011: 3).

An increasing trend among CBA practitioners is to use knowledge-sharing tools to build communities of practice around climate change adaptation. One example is how 20 of the leading climate and development web initiatives decided to set up an informal Climate Knowledge Brokers Group.[9] Another example is how the Climate & Development Knowledge Network (CDKN) has feeds of materials coming from Eldis, Inter Press Service News Agency (IPS) and AlertNet.

Language and culture

Making adaptation tools *available* to local users and stakeholders, such as community-based organisations (CBOs), is only one required step. Making them useful and culturally appropriate is a considerable additional challenge that requires careful tool preparation, sophisticated mediation and a good understanding of local language, knowledge and cultural systems. The challenge concerns the use of language in a tool in both linguistic and technical terms. Most tools are initially developed in English. Some are eventually translated into additional widely spoken languages, such as Spanish, French and Portuguese. Yet these languages are often not spoken in the local communities that are the key target group for CBA. As a result, tool users are faced with the significant challenge of applying the tool in the intended participatory and comprehensive fashion that most CBA-related tools promote. In addition, while a given tool may be targeting a specific user group, the language used in CBA is often couched in complex technical jargon that is not only difficult to understand, but also to translate at the community/local level. Similarly, the tools often need to be adapted to the local cultural context. While a suggested approach might be perfectly culturally acceptable in one setting, it may be very inappropriate in another.

Mismatches in knowledge systems

Top-down approaches to adaptation are often premised on the assumption that modern science can provide a solution to negative climate change impacts that local, traditional knowledge systems cannot. While local knowledge is increasingly being accepted as an integral element of development and most CBA approaches (Huq and Reid 2007), the difference between formal and informal knowledge systems remains a source of conflict. As Huq (2011: 1) states,

> In a world where knowledge equals power, you could be forgiven for thinking that enabling [CBA] boils down to providing local people with information. Conventional approaches to planning adaptation rely on 'expert' advice and 'credible' science from authoritative information providers such as the Intergovernmental Panel on Climate Change (IPCC). But to truly support the needs of local communities, this information needs to be more site-specific, more user-friendly and more inclusive of traditional knowledge and existing coping practices.

Huq (2011) points out that global climate science models cannot accurately predict what will happen in specific locations. Moreover, they generally do not factor-in underlying socio-economic vulnerability in their predictions. So there is an increasing need to combine science with local knowledge, experiences and perceptions of climatic variability and shifts by those most acutely exposed to them.

There are, however, a number of significant challenges in the way that scientific knowledge tends to be conveyed and translated into local languages and made understandable. Many members of vulnerable communities cannot read and understand highly technical scientific information. So it is not enough to simply translate this into local languages – there is a real issue of comprehension and different value systems that also needs to be addressed. Local languages often do not have words for the English equivalents, or communities may be illiterate (Huq 2011). Also, as Berkes (1999) points out, the belief systems that underpin local, traditional knowledge systems tend to differ from the ones that determine 'Western science'. As a result, the former tends to resist the influence from the latter, which represents 'a challenge to the dominant positivist-reductionist paradigm of Western science' which belies the asymmetrical power relations between Western science and local indigenous knowledge.

In response, Robbins (2004: 120) argues that while tools, on the one hand, must 'acknowledge the interested and contextual character of local knowledge', they must also help convey information in an accessible manner to a general audience, for example through audio visual media instead of in written form (Huq 2011).

In this sense, conflicting worldviews are at the heart of the adaptation practice as 'the knowledge of scientific practitioners and other "experts" is imbedded in cultural norms, social relationships, and value-laden judgements, even and especially in large-scale scientific investigations like climate change research' (Robbins 2004: 120). Paying lip service to local knowledge and community participation in community-based adaptation will not rid anyone of the often mismatching knowledge and value systems that are at play between local indigenous peoples and Western science. This challenge will be important to address in tools used for upscaling efforts, as the connections to and involvement of new municipal/sub-regional and national-level stakeholders, who are often more inclined to adhere to Western science than local knowledge, will enter the picture.

Stumbling blocks to converting context-specific case studies into scalable adaptation

The second set of challenges concern the actual process of scaling up documented CBA case studies. First and foremost, there is a need to be clear on what kind of scaling up is aimed for. As highlighted earlier, scaling up can have several meanings. Therefore it is vital to be clear on objectives, scope (vertical, horizontal or a mix of both), how and with whom to carry out this process upfront. For example, scaling up through a large number of new initiatives through a multiplier effect implies moving to scale through high levels of governance, second- or third-tier organisations or through government agencies at the state or federal level. The selection and application of tools will very much depend on which of these parameters are involved. Finally, the tools selection and application will also very much be determined by the desired objectives of ensuring that scaled up and out initiatives are both locally appropriate and of high quality.

One size does not fit all

Empirical lessons show that CBA tools that have been developed for one regional context may need significant modifications before they can be successfully and appropriately applied in another region due to important social, cultural, political and historical differences. For example, CARE's Climate Vulnerability and Capacity Analysis (CVCA) Handbook was primarily developed based on CARE's initial CBA work in the Africa region. Yet, when it was later applied in Latin America, it became very evident that modifications were needed to the suggested approaches and specific Participatory Rural Appraisal (PRA) tools due to important regional differences in governance structures and gender dynamics, among others. Additionally, different local contexts can place different limitations on CBA intervention quality and scale. For example, CBA tools might have to be applied differently in a stable vs. an open conflict context. This also reinforces the notion suggested by Hammill and Tanner (2011) that existing adaptation tools target different niches.

So if one size doesn't fit all, and most tools cater to specific local circumstances, how can recommendations concerning adaptation measures from a given case study be upscaled and adopted by a wider audience? How can we address the specific needs of vulnerable groups (youth, elderly, women, indigenous peoples), who even among themselves tend to be vulnerable in different ways and degrees, while maintaining a broad sense of how these necessarily targeted adaptation measures can be inserted into a broader whole?

Blindness to gender and differential vulnerabilities and capacities

Those most severely affected by climate variability and change tend to be those whose needs and priorities are omitted. It is a known fact that climate change vulnerability, social (and particularly gender) inequalities and lack of voice in such processes are closely linked (UNDP 2007; UNISDR 2011). Large-scale climate change responses, in turn, come with high pressures on budgets and time frames that do not easily lend themselves to great levels of inclusiveness, context specificity and gender-sensitivity. In addition, efforts to upscale CBA in a gender-equitable way will likely be hampered by a gender myth which labels all things domestic, subsistence-level and close to nature as 'female', and progress, technology or science as 'male' (Leach 2007: 68).

For CBA processes to remain context-appropriate and inclusive of the poorest and most vulnerable groups at *any* scale, however, and to avoid negative impacts and perpetuated social inequalities, tools should provide for an in-depth understanding of differential vulnerabilities and capacities to adapt to climate change, and for equitable CBA design. Guidance for both analysis and CBA design needs to instruct how to carry out and address disaggregation of, for example, gender, age groups and ethnicity, beyond community and household levels, and prompt inclusive processes and empowerment of voices too easily unheard.

With few exceptions, such as the UNDP's 'Gender and Community-based Adaptation' guidelines (UNDP 2010) or CARE's CBA toolkit (CARE 2010), existing CBA tools tend to lack meaningful guidance for such a disaggregated understanding of vulnerabilities and capacities, and for inclusive, gender-equitable processes. In particular, most tools tend to be gender-blind[10] or, at best, encourage a superficial treatment of gender. Without a solid understanding of how climate change is affecting different social groups (especially women, children and others with high levels of vulnerability), appreciation of differential adaptive capacity and accompanying capacity building to tackle highly politicised issues of social inequality such as gender, caste and ethnic hierarchies, CBA practitioners will be unable to use tools to develop fair and effective measures at *any* scale (Webb 2011). In addition, given how challenging and complex these processes tend to be in small-scale CBA planning and implementation processes, these challenges will likely be magnified in scaling up and out processes.

The role and importance of institutions

Regardless of scale and the level of planning, CBA planning will predominantly take place at the local level. Yet, in order for scaling up to happen, there is a need to develop the appropriate institutional scaffoldings, along with the necessary linkages and dialogue between the practice and policy levels. The only way for CBA to gain traction and to link up to national and regional policy processes is through articulated institutional scales. Although adaptation practice at the community level can achieve results, local and national institutions will need to be substantially involved to ensure implementation and delivery at different scales through coherent policy, legal and financial frameworks. However, on the flip side, transitioning from local to sub-regional and national policy processes can produce undesirable side effects, such as increased bureaucratic setups and processes, which often tend to empower more the technicians and political actors than those that are living on the front line of climate risks. Similarly, as Biggs *et al.* (2007) have shown, 'some issues and processes are scale-specific and lose meaning when transferred to other scales'. A key challenge is, therefore, how to analyse case studies to identify what constitutes their 'skeleton' or core elements that can be replicated elsewhere vs. elements that will need modification to accommodate different local contexts.

Building and developing institutions and the larger governance structure is thus crucial to scaling up adaptation measures in a policy context in most countries with dispersed and often conflicting interests (Agrawal 2010; Von Korff *et al.* 2010; Kok and Veldkamp 2011). Such structures will be increasingly important to ensure effective CBA, for example to mediate trade-offs between urban consumers or utility companies and rural water consumers in a context of dwindling water availability (Uphoff 1986). For such mediation to produce ethically acceptable outcomes, there is a need to level the playing field, and advocate for the improvement of bargaining positions of the socially and economically excluded, so that the solutions and trade-offs are the result of mutual commitment and inclusiveness (Moser and Ekstrom 2010). Institutions are all the more important as they can also provide long-term stability to these arrangements, enforce new regulations and guarantee fairness and transparency.

How, then, can these different institutional scales be conceived without losing the local empowerment and effectiveness promoted in CBA? Is there, ultimately, a trade-off between quality and scale? To address these questions will require a closer examination of the *function* of institutions at different scales. Different institutional scales are the reflection of a wide variety of territorial and administrative structures, which vary from one country to the next (federal vs. unitary polities). However, the importance of institutional scales is not just in terms of their respective functions, but rather how these scales can be better articulated. The nature of the interactions between all the different key players from both the practice and policy contexts need to be more closely unpacked, as they reflect the subtle differences between cross-scale communications and cooperation, and the peer-to-peer, multi-stakeholder processes that have more to do with the scaling out of local initiatives, through collaborative approaches, coalitions and other horizontal exchanges between CBOs. In this sense, while the replication of successful adaptation approaches at the local level can be achieved by scaling out through horizontal exchanges and capacity development, the longer-term objectives of adaptation are to influence policy decisions, and to mainstream adaptation into national and sectoral plans and programmes. These two different objectives require different processes, and hence different applications of tools and approaches, in order to scale-out on the one hand, in order to achieve a 'critical mass' of best practices, and to scale-up through upstream policy advocacy and institutional reform.

Successful scaling-up of CBA tools will also hinge on increased awareness and heightened social demands for development policies that are consonant with resilience to climate change, while addressing structural issues such as gender blindness. This can be achieved through a variety of means, such as participatory processes and other collaborative planning approaches, which enable multiple stakeholders to share knowledge, develop awareness and improve learning and build capacity for adaptation to take place (Von Korff *et al.* 2010).

To be successfully scaled out and up, CBA tools need to be applied at the correct level, so that best practices in adaptation at the local level can inform regional and national policy. Similarly, national adaptation plans and strategies need to address the local vulnerabilities to climate change identified while applying assessment tools in selected communities. Figure 7.1 illustrates this search for coherence between the practical implications of adaptation at the local level and the national processes of policy advocacy and institutional reform. This

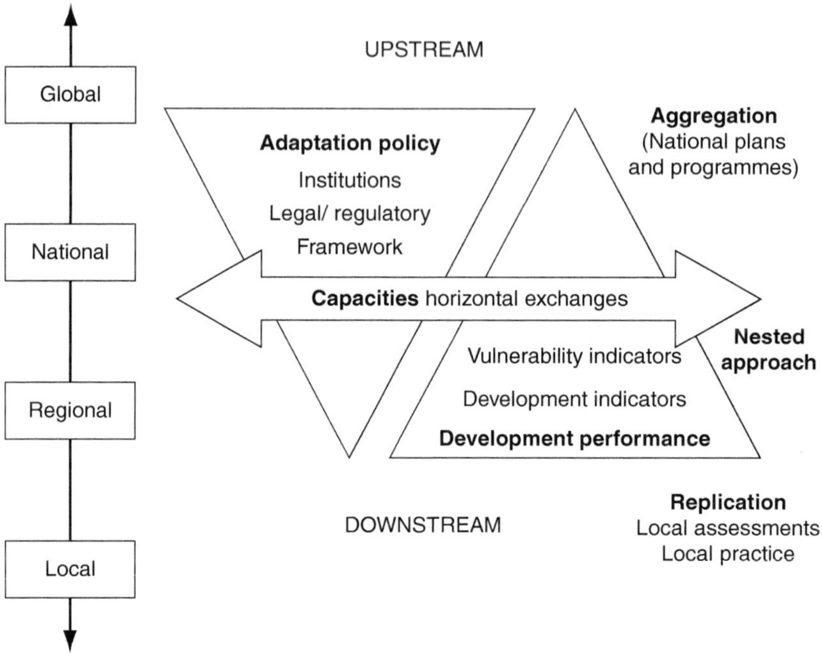

Figure 7.1 Scaling up adaptation tools and approaches (source: based on Brooks *et al.* 2011).

twin-track approach, recently suggested by Brooks *et al.* (2011: 24) for evaluating adaptation programmes, identifies a first track that tends to seek upstream effects on climate risk management through policy advocacy and the development of institutional capacity for broader implementation at the national, regional and local levels. A second track addresses the downstream effects of adaptation in terms of its ability to improve development performance by reducing vulnerability and making livelihoods more resilient.

If applied to scaling up adaptation practice, this twin-track scheme enables local CBOs to move to scale through nested or network approaches, sharing lessons learned, pooling capacity development and facilitating access to technology through multi-stakeholder platforms. The coupled development/vulnerability indicators also help to keep tabs on both development performance and the impact of climate hazards on development assets at the community level. The horizontal exchange of tools and practices, proposed under this framework, are critical for the replication and scaling out of best practices. Local adaptation practice is conveyed by practitioners themselves and provides a time-tested means to increase the adoption of new tools and approaches.

On the other hand, these multi-stakeholder platforms and networks of local practitioners can aggregate their efforts to better influence and inform national policy, supporting policy implementation and capacity development. The central role of capacity development, shown in Figure 7.1 as the cross-track arrow, enables building bridges between policy and practice, thus minimising the trade-offs between scale and quality.

Coming to terms with reality

Adaptation policy will require not only a range of appropriate tools and methods to assess existing and future vulnerability, but will also require robust institutions to follow through with adaptation measures. Adaptation measures are typically quite complex and designed for the 'ideal' situation. Community adaptation plans are most often geared to protect the most vulnerable segments of society, and seek to deliver on these objectives against high ethical standards. However, many well-intentioned policies often come up against stark realities and are frequently the victim of special interest groups and other more powerful stakeholders (Oliver-Smith 2001). Getting adaptation practice off the ground and mainstreamed will require significant investments in institution building and staff capacity development, which both require significant time and resources (both human and financial). For many involved actors (e.g. local government and vulnerable communities) it may simply not be possible to demand such a high standard of work on a broader scale.

In light of these resource constraints, after a given 'model' of any CBA-related work has been created, what essentials of good CBA practices can be identified for replication more broadly? This question also links with the issue of higher-level policies and what standards and incentives need to be put in place for CBA to take off.

Conclusion

CBA practice has grown rapidly over the past years, and today it is a crowded field where competing approaches and the proliferation of tools tend to confuse, particularly its end users at the local level. In the context of the broader debate on ways forward on scaling up adaptation to climate change, and of a small but growing number of reviews of tools and methods for adaptation generally, this chapter has sought to discuss why scaling up presents a particular challenge for tools for CBA. Following a very brief overview of adaptation tools reviews and stock-taking efforts, and a discussion of what scaling up means in the context of CBA, we discussed two main sets of challenges we perceive as crucial: (1) building a knowledge base on CBA tools and approaches that can be shared and scaled up; and (2) how to go about scaling up CBA practice, through which approaches and institutional processes.

This chapter identifies specific challenges associated with building and sharing knowledge on scalable CBA experiences. It also points to a second set of challenges which are related to the very process of scaling up CBA itself. Overall, we discussed both intrinsic limitations of the contexts in which tools are applied (e.g. capacity, context, gender blindness, cultural resistance) and the complexity of the broader institutional processes (e.g. governance, role of institutions) involved in scaling up CBA.

While CBA practice has made considerable progress in addressing issues of capacity building, gender equity and participatory approaches, there are still many outstanding issues concerning the governance and institutional dimensions required for the scaling up of CBA tools to really take place and become mainstreamed. Our chapter argues that for this to happen there is a need to tailor existing and new tools to the broader development context, while maintaining a robust score-keeping of development gains and vulnerability-related losses.

By linking policy and practice through learning, these tools can become a way of not only extracting information on adaptation performance and vulnerability scenarios, but more importantly they can contribute to better accountability and thus enable *real* replicability. Considerable amounts of resources will be invested in adaptation tools over the next decades, and we need to maintain realistic expectations regarding their application in terms of what we can reasonably achieve through adaptation, and what needs to change in terms of broader development policy for replication to happen. This can in part be achieved through the establishment of platforms and clearing-house mechanisms to allow learning on adaptation to happen at a pace that is really needed.

Ultimately, the test will be whether the gamut of existing CBA tools enables an understanding of underlying drivers of risk in the face of climate change and facilitates the identification of appropriate strategies to address them. It is important to remember, however, that tools can only be as enabling as the environment that supports them. Without strengthening good governance and community empowerment, tools for scaling up CBA will remain without tangible impact on people's resilience.

Notes

* The authors wish to thank Richard Ewbank, Fiona Percy, Sarah Wiggins and Tom Tanner for their substantive contributions to this paper.

1 Tine Rossing is an independent consultant. Agnes Otzelberger and Pascal Girot both work for CARE International as members of its Poverty, Environment and Climate Change Network.

2 For example, CARE's Climate Change Vulnerability and Capacity Assessment (CVCA) Handbook, the Community-Based Risk Screening Tool – Adaptation & Livelihoods (CRiSTAL), the E-Learning Tool for Planning for Community-Based Adaptation to Climate Change, or Christian Aid's Adaptation Toolkit – Integrating Adaptation to Climate Change into Secure Livelihoods.

3 It should be noted that some tools, such as CARE's Climate Change Vulnerability and Capacity Assessment Handbook, clearly state that it is a tool for assessment only.

4 This definition was expressed by Saleemul Huq, Senior Fellow of the International Institute for Environment and Development (IIED), during the CBA5 conference in Dhaka 2011. See 'Scaling up community-based adaptation (CBA) to climate change', posted by Africa Press International, 26 April 2011.

5 These definitions were culled from definitions outlined by Snapp and Heong (2003) and expressed by Saleemul Huq, Senior Fellow of the International Institute for Environment and Development (IIED), during the CBA5 conference in Dhaka 2011.

See 'Scaling up community-based adaptation (CBA) to climate change', posted by Africa Press International, 26 April 2011.

6 This argument is partly supported by Mearns (2011).

7 See www.acicafoc.org and, more specifically, www.wix.com/z20403/cosecha-de-agua-de-lluvia#!.

8 www.careclimatechange.org/tk/cba/en.

9 This was one of the key outcomes from the Climate and Development Knowledge Management Brokers Workshop held in Eschborn, Germany, 3–5 June 2011. This was co-hosted by GIZ and PIK-Potsdam. Twenty of the leading climate and development web initiatives got together to discuss how they could better collaborate. The group included well-established players like the Adaptation Learning Mechanism (ALM), Eldis and Africa Adapt.

10 Neglecting the differential needs, roles, priorities and skills of women, girls, men and boys at different life stages.

References

Agrawal, A. (2010). The Role of Local Institutions in Adaptation to Climate Change. Paper commissioned by the World Bank Group for the Social Dimensions of Climate Change workshop.

Barnard, G. (2011). Seeking a Cure for Portal Proliferation Syndrome. 8 June. http://cdkn.org/2011/06/portal-proliferation-syndrome.

Berkes, F. (1999). *Sacred Ecology: Traditional Ecological Knowledge and Resource Management*. Philadelphia: Taylor and Francis.

Berkes, F. and D. Jolly (2001) Adapting to Climate Change: Social-Ecological Resilience in a Canadian Western Arctic Community. *Conservation Ecology*, 5 (2), 18.

Biggs, R., C. Raudsepp-Hearne, C. Atkinson-Palombo, E. Bohensky, E. Boyd, G. Cundill, H. Fox, S. Ingram, K. Kok, S. Spehar, M. Tengö, D. Timmer and M. Zurek (2007). Linking Futures Across Scales: A Dialog on Multiscale Scenarios. *Ecology and Society*, 12 (1), 17.

Birkmann, J., G. Tetzlaff and K. Zentel (2009) *Addressing the Challenge: Recommendations and Quality Criteria for Linking Disaster Risk Reduction and Adaptation to Climate Change.* Bonn: DKKV.

Brooks, N., S. Anderson, J. Ayers, I. Burton and I. Tellam (2011) *Tracking Adaptation and Measuring Development*, London: IIED.

Burton, I. (2011) Closing Keynote Speech at the CBA5 conference in Dhaka, Bangladesh, 31 March.

CARE Community-based Adaptation Toolkit, Digital Toolkit, version 1.0 (2010) Produced by CARE International with technical inputs by the International Institute for Sustainable Development (IISD). [online] www.careclimatechange.org/tk/cba/en.

Ewbank, R. (pers. comm.), RE:1st rough draft of CBA Book Chapter on Upscaling Tools for Community-based Adaptation – Emerging Lessons, 9 August 2011.

Gambarelli, G. (2011) Presentation on Climate Change Adaptation Vulnerability Assessments using Adaptation Tools (CRiSTAL and CVCA) in East Africa and Central America: Next Steps, CBA5 conference session on tools, Dhaka, Bangladesh, 29 March.

Glantz, M. (1988) *Societal Responses to Regional Climate Change: Forecasting by Analogy*, Charlotte, NC: Westview Press.

Glantz, M. (1990) *ENSO Teleconnections Linking Worldwide Climate Anomalies: Scientific Basis and Societal Impact*, London: Cambridge University Press.

Glantz, M. (2003) Usable Science 8: Early warning Systems: Do's and Don'ts, Workshop Report, 20–23 October, Shanghai, China, Colorado: UCAR. http://ccb.colorado.edu/warning/docs/report.pdf.

GTZ (2009) Workshop on Mainstreaming Adaptation to Climate Change Guidance and Tools. GTZ House Berlin, Potsdam Square 28–30 May. www.gtz.de/de/dokumente/en-climatemainstreaming-adaptation-workshop-report.pdf.

Hammill, A. and Tanner T.M. (2011) Harmonising Climate Risk Management? Adaptation Screening and Assessment Tools for Development Co-operation. Working paper for the OECD Task Team on Climate Change and Development Co-operation, May.

Huq, S. (2011) Improving Information for Community-based Adaptation, Opinion: Lessons from Adaptation in Practice, IIED, October. http://pubs.iied.org/pdfs/17103IIED.pdf

Huq, S. and H. Reid (2007) *Community-based Adaptation: A Vital Approach to the Threat Climate Change Poses to the Poor*, London: IIED.

Kok, K. and T.A. Veldkamp (2011) Scale and Governance: Conceptual Considerations and Practical Implications. *Ecology and Society*, 16 (2), 23. [online] www.ecologyand-society.org/vol. 16/iss2/art23.

Larsen, R.K., A.G. Swartling, N. Powell, L. Simonsson and M. Osbeck (2011) A Framework for Dialogue between Local Climate Adaptation Professionals and Policy Makers: Lessons from Case Studies in Sweden, Canada and Indonesia. Research Report, Stockholm Environment Institute (SEI).

Lavell, A. (2004) Local Level Risk Management: From Concept to Practice. CEPREDENAC-UNDP.

Leach, M. (2007) Earth Mother Myths and Other Ecofeminist Fables: How a Strategic Notion Rose and Fell. *Development and Change*, 38: 67–85.

McCarthy, J.J., O.F. Canziani, N.A. Leary, D.J. Dokken and K.S. White (eds) (2001) *Climate Change 2001: Impacts, Adaptation, and Vulnerability. Contribution of Working Group II to the Third Assessment Report of the Intergovernmental Panel on Climate Change.* Cambridge: Cambridge University Press.

Mearns, R. (2011) Scaling Up Community-based Adaptation. 5 April. http://blogs.world-bank.org/climatechange.

Mitchell, A. and A. Otzelberger (2010) Review of field based risk management tools. Unpublished report.

Moser, S.C. and J.A. Ekstrom (2010) A Framework to Diagnose Barriers to Climate Change Adaptation. *Proceedings of the National Academy of Sciences*, 107 (51), 22026–22031.

Olhoff, A. and C. Schaer (2010) *Screening Tools and Guidelines to Support the Mainstreaming of Climate Change Adaptation into Development Assistance: A Stocktaking Report*. New York: UNDP.

Oliver-Smith, A. (2001) Displacement, Resistance and the Critique of Development: From the Grass Roots to the Global, Final Report Prepared for ESCOR R7644 and the Research Programme on Development Induced Displacement and Resettlement, Refugees Study Centre, Oxford University.

Percy, F. (pers. comm.). RE: GPF and CBA, 24 November 2011.

Ribot, Jesse C. (2009) Vulnerability Does Not Just Fall From the Sky: Toward Multiscale Pro-poor Climate Policy. In: Robin Mearns and Andrew Norton (eds), *Social Dimensions of Climate Change: Equity and Vulnerability in a Warming World.* Washington, DC: World Bank, pp. 75–102.

Ricoy, A. (2011) Presentation on FAO's e-Learning tool at Planning for Community-Based Adaptation to Climate Change, CBA5 conference session on tools, Dhaka, Bangladesh, 29 March.

Robbins, P. (2004) *Political Ecology: A Critical Introduction*, London: Blackwell.

Schipper, L. (2009) Meeting at the Crossroads? Exploring the Linkages Between Climate Change Adaptation and Disaster Risk Reduction. *Climate and Development*, 1 (2009), 16–30.

Snapp, S.S. and K.L. Heong. (2003) Scaling up: Participatory Research and Extension to Reach more Farmers. In Pound, B., S.S. Snapp, C. McDougall and A. Broun (eds) *Uniting Science and Participation: Managing Natural Resources for Sustainable Livelihoods*. London: Earthscan.

Tanner, T.M. and B. Guenther (2007) *Sharing Climate Adaptation Tools: Improving Decision-making for Development*. Report of Geneva Workshop, 11–12 April, Institute of Development Studies, Brighton. www.iisd.org/pdf/2007/sharing_climate_adaptation_tools.pdf.

UNDP (2010) *Gender, Climate Change and Community-based Adaptation: A Guidebook for Designing and Implementing Gender-sensitive Community-based Adaptation Programmes and Projects*. New York: UNDP.

UNISDR (2011) Global Assessment Report on Disaster Risk Reduction 2011: Revealing Risk, Redefining Development. UNISDR. www.preventionweb.net/english/hyogo/gar/2011/en/home/index.html

Uphoff, N. (1986) *Local Institutional Development: An Analytical Sourcebook with Cases*, Hartford, CT: Kumarian Press.

Von Korff, Y., P. d'Aquino, K.A. Daniell and R. Bijlsma (2010) Designing Participation Processes for Water Management and Beyond. *Ecology and Society*, 15 (3): 1. [online] www.ecologyandsociety.org/vol. 15/iss3/art1.

Webb, J. (2011) Presentation on practical tools and resources for CBA: CARE's toolkit and CBA standards, at the CBA5 conference session on tools, Dhaka, Bangladesh, 29 March.

8 Agriculture and climate forecasting

Sudip K. Rakshit, Richard Ewbank and Dinanath Bhandari

With agricultural productivity fundamentally dependent on the four inter-related factors of soil quality, solar radiation, the occurrence of pests and diseases and the prevailing climate, meteorological and climate forecasting are essential tools for the farmer. Short-term and seasonal forecasts communicated through accessible channels and in a timely fashion will, if blended with information about the agricultural calendar and relevant practices and technologies, enable farmers to make critical decisions about on-farm management that can either mitigate the impacts of negative conditions or enable them to take advantage of favourable ones. Farmers in developed countries typically use multiple information channels to access meteorological information on timescales from the next 24 hours to the next season to guide management decision-making. In addition, information on the likely impacts of climate change and long-term climate forecasts is increasingly deployed to assist enterprise planning.

However, small-scale farmers in developing countries rarely receive climate information in either a format that enables practical application or at a time that facilitates its application to agricultural decisions. The reasons for this vary widely but include the comparatively low level of resources available to national meteorology agencies to generate the fine-resolution, location-specific forecasts that farmers need – for example, Africa has only one-eighth of the minimum WMO recommended number of climate stations; a decline in funding from both governments and donors in agricultural extension systems that could act as intermediaries for forecast information; and when forecasts are disseminated, they typically lack an explanation of the uncertainty they contain and are unrelated to agricultural practices and processes.

Topographic variation creates particular challenges, given the functions of altitude and slope affecting exposure and wind circulation. Farmers in upland areas often need more specialised information, a situation that warrants greater investment in meteorological stations and the long-term management of information generated from them. Hydrological systems connect catchment areas with those downstream, so climate in higher areas has significance to lower altitudes, especially where management of irrigation resources is concerned. Many mountainous areas lack the adequately distributed weather stations needed to get analysable and useable data. For example, due to altitude varying from below 100 m to 8,848 m above sea level, Nepal harbours five major climate regions and several unique micro-climates that are yet to be fully understood, all within a range of 200 km. The country has only 166 weather stations that have provided analysable data since 1976, but these do not represent the country's diverse climate (see Table 8.1). Out of 166 stations, 34 have both temperature and rainfall data, 132 have rainfall only and ten stations only temperature data. Areas above 4,000 m occupy more than 10 per cent of the surface area, but contain no weather stations. Consultation with local communities revealed that they lack access to the information from these stations (Gurung and Bhandari, 2008; Practical Action Nepal, 2010a).

In the absence of scientific information, small-scale farmers and pastoralists have over generations developed their own systems of local knowledge and predictive practice. This includes historical knowledge and categorisation of seasons that may well differ from that used by the conventional meteorological community, as well as a variety of indicators, often based on the behaviour of birds, insects and plants, to predict some of the most important events on the agricultural calendar, such as the onset of rainy seasons, the likely reliability of the rains through the growing season and the imminence of rains during the season.

Traditional climate prediction may be left to elders, families or clans that specialise in this skill. Some of this knowledge may be unique to particular areas, depending on local environmental and socio-cultural settings, while other indicators, such as bird migration patterns, may have relevance and scope for wider use. There is debate about which of these skills or indicators are useful or have a basis in science that can be exploited. Banerjee *et al.* (2003) concluded that climate-related folklore in West Bengal had no scientific basis. On the other hand, close correlation between bird migration patterns and the movement of the intertropical convergence zone (ITCZ) has been researched in West Africa where, for example, the grey-headed kingfisher moves north with advancing monsoon rains in April and May before migrating back south in October (Cheke and Tratalos 2007).

Table 8.1 Weather stations across Nepal

Station number	*Region*	*Altitude (metres above sea level)*	*Number of stations measuring temperature*	*Number of stations measuring rainfall*
1	Tarai	60–200	11	30
2	Shiwalik	200–1,000	9	47
3	Middle hills	1,000–2,500	22	80
4	High hills	2,500–4,000	2	9
5	Mountain	>4,000	0	0
	Total		44	166

Source: Adapted from Department of Hydrology and Meteorology, Nepal (2009).

Applying scientific forecasts to agricultural decision-making

Where attempts have been made to increase the use of forecasts by building on local participation, results have been encouraging. In Zimbabwe, work to support farmers' access to seasonal forecasts from 2000 to 2004 used participatory climate forecast workshops to assist farmer groups to better understand the forecast and apply it to their farm management decisions (Pat *et al.* 2005). The process began with farmers reflecting on the past season's rainfall, whether it agreed with the official seasonal forecast and how this related to the success or otherwise of their agricultural management practices. This was followed by combining farmers' local climate indicators for the next season with the scientific forecast to get a downscaled seasonal forecast more relevant to their location, including an assessment of El Nino and its potential influence. The final stage involved farmers and the local agricultural extension service officer assessing appropriate farm management practices for the coming year, taking into account the forecast, the local indicators and seed availability.

This use of forecasts was associated with an increase in yield, compared with farmers' typical range of harvests, of 9.4 per cent across the first two years and 18.7 per cent in the 2003/2004 season. An interesting feature was that the observed effect was both greater in magnitude and more significant in the second year, which suggested that forecast value may be higher in La Nina years. So although it may be in drought (El Nino) years that forecasts are of the greatest value to national-level planners preparing for food insecurity, forecasts seem to benefit farmers most when they give them the opportunity to take advantage of good conditions.

Similar work in Mali has also shown the value of forecasts. Since 1982 Mali's National Meteorological Service has supported 2,500 farmers to increase their access to meteorological information in order to better manage climate risk (Helmuth *et al.* 2010). Initially this involved working with a small group of farmers in the southern Bancoumana region, each managing two plots, one using local indicators and the other on which management decisions were made using agro-meteorological information. Farmers received rain gauges and training on how to use measurements with the seasonal calendar to make choices on crop variety and planting dates. They also received ten-day bulletins on hydrological, meteorological, agricultural and pest conditions in addition to daily and three-day weather forecasts. Demand has subsequently spread the programme to six districts, with a focus on farmers defining the climate-related data and products they need, which are then provided by a number of agencies, including the Ministry of Agriculture's extension services, the media and the Meteorological Service.

More recently this has included expanding the approach to an increasing number of farmers' organisations, rural programmes, development agencies and NGOs. Roving seminars have trained over 500 farmers to increase their knowledge on the effects of weather and climate change on rural activities, distributing rain gauges and developing a core group of farmers to collect and report weather and climate data. For the 2010 season, crop yields for farmers taking management decisions with agro-meteorological information compared to national averages for millet, sorghum and maize were 36 per cent, 37 per cent and 48 per cent higher, respectively. Similar results for millet and sorghum for the 2003/2004 season were 44 per cent and 10 per cent higher, respectively, than on fields cultivated without agro-meteorological advice.

Communicating climate and climate change knowledge to farmers has used the farmer field school structure as a vehicle for training and information dissemination. Climate field schools (CFSs) were first developed in Indonesia (Boer *et al.* 2003) modelled on the farmer field schools designed to promote integrated pest management. The basic objective was to develop an effective method for communicating climate forecast information to end users (in this case, farmers). More specifically, CFSs were intended to (1) increase farmers' knowledge of climate and their ability to anticipate extreme climate events; (2) assist farmers in observing climatic parameters and their use in guiding farm activities; and (3) assist farmers to interpret climate (forecast) information, in particular for planting decisions and cropping strategy. CFS development followed a two-stage process. The first 'socialisation' phase covered eight months and focused on increasing farmer knowledge on climate and the use of seasonal forecast information to develop a cropping strategy. The second 'institutionalisation' phase covered a further 32 months and focused on putting this strategy into operation and capacity building farmer groups to integrate climate and forecast information into their farming activities. In order to prepare for the implementation of these two stages, agricultural extension workers were trained by Meteorology Department staff to act as intermediaries/trainers. A number of modules were developed and field tested, including elements of weather and climate – including the difference between weather and climate; rain formation processes; understanding terminology used in seasonal forecasting; understanding probability concepts; use and calibration of non-standard weather/climate measurement tools; use of climate forecast information in planting strategies; using water balance concepts to estimate irrigation requirements and flood risk and quantifying the economic benefits of using climate forecast information.

CFSs took a strongly participatory 'learning-by-doing' approach, with farmers putting module information into practice over the agricultural season and reflecting on their experiences through a continuous process of group discussions and analysis with extension staff to inform subsequent action and strategy revision. When the process was evaluated, 78 per cent of farmers felt that their ability to integrate climate and forecast information into their cropping strategies had increased significantly (7/10 or better). The key implementation challenges identified were: translating climate information into user-friendly language for farmers; and integrating this into effective adaptation.

Using both scientific and local knowledge for climate prediction

The Climate Change Adaptation in Africa programme (CCAA) has expanded the focus on seasonal forecast provision through eight projects across the continent (Ziervogel and Opere 2010). These have explicitly sought to integrate local or indigenous knowledge with scientific seasonal forecasts, recognising the challenges involved in bringing two very different traditions of forecasting together. In Western Kenya, for example, this involved bringing together scientists from the Kenya Meteorological Department with the Nganyi, known for their expertise and knowledge of local climate indicators. The aim was to develop a consensus seasonal forecast that would then be communicated to the wider community incorporating information from various government departments in the form of advisories covering health, agriculture and so on. Five seasonal forecasts have been communicated in this way and have led to the allocation of resources for a resource centre to harmonise local and scientific forecasting and further its use for local application.

In Nepal, Practical Action analysed meteorological data available across the country to identify climate hot spots (Practical Action Nepal 2010b). These hotspots were later assessed through community-based tools to understand whether and how local communities have experienced climatic changes and impacts and to investigate the role of local knowledge in forecasting climate (Practical Action Nepal 2010a). Local communities did not have access to the meteorological information, and in the absence of scientific information, communities used current weather phenomena to forecast the upcoming season based on their past experiences. While the behaviour of birds and insects are used to predict short-term changes, specific events are generally used to predict changes in future seasons. So when communities in the Gandaki river basin witness the Siberian crane flying north in March, this migration indicates the time for seasonal movement to higher, hilly areas to sow seeds. Similarly, when cranes fly south in September/October, they indicate the beginning of winter and the time for harvesting summer crops and returning to the lower hills. The timely migration of these cranes provided important information to communities before there was a formal calendar. Other phenomena include the appearance of red clouds in the evening during August, which are used as an indicator that the rains would end within a week. Conversely, the same red clouds in the morning in April denote rainfall arriving within a week, and so act as a signal to prepare for sowing maize in upland areas. Likewise, the occurrence and behaviour of black clouds are used to estimate the imminence of rainfall.

Longer-term indicators include lightning in the midsummer monsoon (June–July), taken as an indicator of bad weather in the coming seasons. More intense rainfall in July indicates the possibility of drought in the following season, and large raindrops in January rainfall indicate a likelihood of poor pre-monsoon rains and hailstones in the approaching monsoon season, which can have severe impacts on agricultural production (see Box 8.1). Reduced rainfall in winter

Box 8.1 The impact of extreme events

Hukum Singh Gurung (59), a resident of Khetbari village, Chitwan, had managed his vegetable farm with a drip irrigation system to cope with the limited water resources available. In April 2007 there was a huge hailstorm in the village. His drip irrigation pipe was severely damaged, along with his zinc plate roof. 'I had never seen such big and damaging hailstones before', he explained sadly. Apart from the cost of repairing his pipe (about $31), he lost his whole crop of vegetables that would have generated about $300 – his family budget for more than six months. This story was repeated by all the farmers in the area. 'It's no use how favourable other factors are on other days; a single event is enough to cause a great loss within a short time', says Janga Sarki (45). Earlier in the year there was a dry wind storm which uprooted hundreds of trees in their community forest. Subsequently, there were only two rainfall events between October 2007 and February 2008; one in January and another in February. Both brought hailstones and caused widespread damage to winter vegetable crops in the area. 'I had never seen such a big heap of hailstones before' said Sher Bahadur Tamang (50) of nearby Bhotedhap village.

indicates windstorms in April/May, which households take as a signal to focus on roof maintenance. These forecasts are based on the past experience of similar events in the locality. Extreme events are considered particularly useful in predicting weather events, and the trend of past extreme events is used to describe and forecast future weather trends.

These local indicators and climate knowledge are passed down from elders to younger generations, incrementally building on memories of past lives. For example, if a weather event takes place, such as sudden unexpected rainfall encountered while carrying a back-load home, the affected family member would share how she or he encountered the event and what the consequences were. Then, other members, and particularly elders, would recall similar events and the consequences they have had in the past. The family would discuss the trends connecting these events and, probably with friends while working on the farm or during grass and fuelwood collection, exchange ideas. Family members, particularly women, would share this information while visiting parents. Specific community members have particular information such as the *Chepangs*, who focus on indicators such as bats and the *chiuri* (or butter nut tree), and traditional healers, who use medicinal plants to gauge weather trends. These sources of knowledge have combined to provide intuitive perceptions of future climate for their communities.

Where past meteorological data are not available for specific climate areas, local indicators recognised by the community can be used as proxy indicators of the impacts of climate change that are happening in the locality, as well as the sensitivity of agriculture. Combined information from both meteorological and community experience provides guidance to devise future coping and adaptation strategies. In East Rapti watershed, information from meteorological observation showed that rainfall was increasing but hydrological monitoring indicated river discharge was decreasing. Adding to this apparent contradiction, the community confirmed that they were increasingly experiencing both extreme situations of flood and drought. Although average rainfall had increased, the degraded watershed could not absorb the more erratic rainfall pattern, resulting in increased flood events; the lack of proper recharge later decreased stream flows.

Similarly, in the Jugedi Stream watershed, more intensive and erratic rainfall has worsened soil erosion, reducing both water availability and agricultural production. Agricultural practices such as slash-and-burn cropping in steep slope areas have degraded the watershed, aggravating the drought situation during drier-than-usual monsoon seasons and increasing the risk of landslides and floods during normal or wetter seasons. Combined analysis of information from both local indicators and meteorological data has helped attribute the root causes and consequences of slash-and-burn practices, helping farmers understand the factors leading to their increased vulnerability. Improvements in cropping practices, better soil conservation and utilisation of water resources and the conservation of forests in the catchment area have enhanced ecological conditions, reducing landslides despite the more intense rainfall experienced. Communities have more precise information on local phenomena. Combining this with meteorological observations has helped farmers better understand both causes and consequences.

In Africa, studies on the use of indigenous knowledge by the Maasai in Tanzania (Kilongozi *et al.* 2005) demonstrated a careful assessment of the reliability of their local forecasting methods, which leads to a combined use of both scientific forecasts and local indicators to increase skill. The prevalence and pragmatic use of informal and formal climate prediction practices demonstrates the interest of small-scale farmers and herders in the use of climate information, combining forecasts from different sources to triangulate information and a skill in navigating its inherent uncertainty.

Work by INADES in central Tanzania, supported by Christian Aid's Climate Change Innovation Fund, was similarly based on these concepts of combining local knowledge with available meteorological information (Ewbank 2011). The first step was to carry out a participatory assessment of local perceptions and knowledge on climate and weather forecasting. The priority climate concern expressed by community members was drought, which was assessed to have increased, especially in the last five years, and was given more high-priority designations than any other factor in the assessment exercises. Although there was some reference to reduced amounts of rainfall, more emphasis was placed on its increased unreliability – this included more variable starts to the rainy season; rainfall in more condensed episodes and with less reliability in both timing and where in the area rain was likely to fall in any one event (e.g. in 2008/2009, only March had reliable rainfall); and an expanding mid-season dry spell that has extended from its usual time in February back into January.

Other climate risks cited included an increased occurrence of strong winds that cause dust storms, increase soil erosion and destroy housing. Communities associated both changing rainfall and the increased occurrence of strong winds with local deforestation as much as with what they have heard about climate change and greenhouse gas emissions. They clearly attributed the beginning of changes in climate with the expansion of farmland at the expense of forests. Spontaneous change to agricultural practices have been limited to some increase of sorghum, sunflower, chick peas and onions at the expense of maize and beans.

Farmers use both local indicators as well as seasonal forecasts on which to base agricultural decisions. Seasonal forecasts are obtained from the radio in the first instance, with some access through the District Agricultural Office to the Ward Extension Officer, who distributes information through shops and schools in September/October. However, these forecasts lack more locally specific information and local knowledge is generally regarded as the most dependable. Local indicators include the behaviour of certain animals and insects, the migration patterns of certain bird species[1] and the appearance of certain stars. Villages typically reported 15–17 different types of local indicator (see Table 8.2), with these focusing on the way a rainy season is likely to start, whether it will be a good year or a drought.

Although there is some concern that local indicators are not performing as well with the increasingly erratic climate, the communities still regarded them as more reliable, and they performed better in recent El Nino events than the less-specific meteorological information they had received. For instance, for the 2008/2009 season, the seasonal forecast was for good rains, but in the event it was a poor season with erratic and reduced rainfall. Local indicators performed better in this instance. They tend to focus on the coming season, with a particular emphasis on the likely start date for the rains and whether it will be a good or bad season. Farmers clearly felt that combining both scientific and local information so they can see where these agree greatly strengthened the value of seasonal forecasts.

Following the analysis of climate, the project carried out climate change vulnerability assessments to identify specific measures that communities could take. These included all sections of society, and some efforts were made to promote equal gender representation (63 per cent of participants were men, 37 per cent women) and the involvement of youth. Vulnerability assessment involved understanding livelihoods and livelihood options; understanding risk, vulnerability and coping strategies (including those identified in the participatory assessments and who in the community were most vulnerable); developing viable adaptation options for managing potential risks associated with climate change; and finally assessing the current climate risks and institutional frameworks. This led to action plans covering the next three years to implement adaptation activities, covering issues such as introducing new drought-resistant maize and sorghum varieties, digging ponds to harvest water and to use for irrigated horticulture, vaccinating livestock to tackle increased disease outbreak and tree planting to reduce soil erosion and improve local climate.

In all four villages the development of the vulnerability assessment led to the formation of farmer field schools to carry out on-farm research into drought-resistant agriculture. The focus on drought-resilience was evident in the selection

Table 8.2 Local climate indicators

Type	*Behaviour*	*Timing*	*Indicator of*
Trees	Heavy flowering of baobab, mango and some other species	August–October	Drought conditions
	New leaf formation in mihingili, mtindimbuli and other species	September–October	Time to start land preparation, rain approaching
	Leaves turn yellow in baobab and other species	August–September	Drought conditions
Animals	Appearance of pangolins	August–October	Good rains
	Elephants appear in the local area	September–November	Time to start land and seed preparation
	Python appears in the village	November	Likelihood of a good harvest
Insects	Butterflies swarming	February–April	Drought conditions
	Cobwebs cover the baobab trees	September–October	Drought conditions
	Termites move into grazing areas	September–October	Drought conditions
Birds	Large ground bird (probably ground hornbill) starts calling	September–October	Farm preparation and a good rainy season
	Hondohondo (quelea) bird migrates from west to east in swarms	September–October	Farm preparation and a good rainy season
	Mguulo bird starts calling	August–October	Depending on the call, either good rains approaching or drought
	The sangalujembe bird (white with black wings) appears	October–December	Drought conditions
Stars	One of the two stars that appear to the north shines brighter	July	Good rains
Wind	Heavy red dust wind (dust devils) blowing from east to west	September–October	Time to start land and seed preparation

of issues to research, which included testing new short-season maize and sorghum varieties, the use of four different tillage options and use of manure. The two villages in Manyino District are more drought vulnerable and have focused on sorghum, while both maize and sorghum have been trialled in Chamwino District. In both areas, crops using manure and the magoye ripper to break up subsoil pans and allow both moisture and plant roots to penetrate deeper tended to perform better, although in one area, Mbwasa, localised drought had prevented trials from starting.

Other measures have included the establishment of rain gauges in all four villages. These have been manufactured locally as the standard gauge cost TSh 2 million, which is unaffordable by a farmer group. The local gauge cost TSh 250,000, just one-eighth of the cost. Dodoma Meteorology Station staff have provided training on management, record keeping and design – the initial gauges needed a sharper lip to increase the accuracy of rainfall measurement before they could be registered. The rain gauge committees submit their data to the Dodoma station as well as use it for their own crop management purposes.

The community highlighted ten major impacts of the project, these broadly splitting into two groups – those that have enhanced the capacity of the community to understand climate and manage climate risks, and more practical measures that have enabled these risks to be addressed. Those in the former category include understanding climate change and knowing how to respond, understanding how to integrate this into local planning and better access to climate advisory services. More practical measures include the increased use of manure, access to drought-resistant crop varieties and a greater understanding of the benefits of various tillage methods.

Although farmers assessed the increased use of manure as the most important impact, they considered the increase in their own capacity for understanding climate change and planning resilience to have made the most progress with both combining local and scientific methods and understanding climate change and knowing how to respond improving from a particularly low level before the project (see Figure 8.1; dark grey shows the situation before the project, light grey after the project). Issues where communities felt they had made progress but could make more were primarily better planning to integrate climate change and more adoption of new technologies, including exchanging experience with other farmer field schools. This highlights the value placed on farmer-to-farmer linkages in building resilience. It also reflects the farmers' view that having developed a link with meteorological science, they see considerable potential in improving their access to and use of better-quality information.

Conclusions and recommendations

Increasing the local relevance of scientific forecasts

An important first step is to increase the capacity of national meteorology services to generate finer-resolution, more location-specific forecasts. Most developing countries have a significant deficit in the grid of climate stations and measuring points that would enable this, as well as a capacity to develop and support the communication channels and required networks of intermediaries that are needed. The scepticism on the part of farmers regarding forecasts is based partly on this lack of local specificity, but also the need to understand and know how to apply probabilistic information to agricultural decision-making.

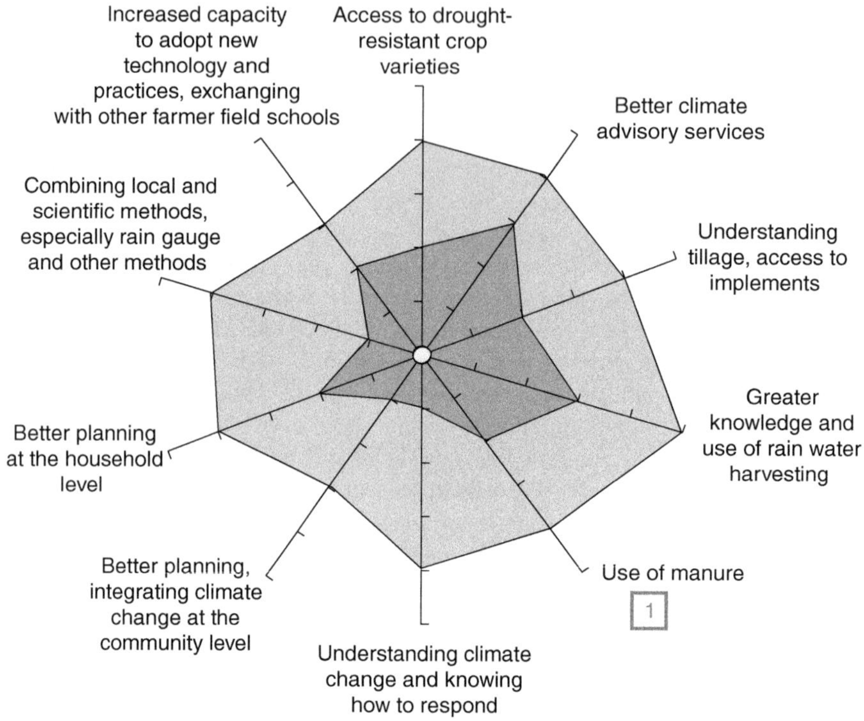

Figure 8.1 Group assessment of project benefits and the impact compared to the situation before the project.

Increasing the science behind local knowledge

Farmers frequently express concern that their local indicators have become less efficient as they have experienced changing climate trends. However, as the work in Tanzania has shown, when they realise that there is a scientific basis behind an indicator they have been using, their confidence in its application increases. This enables farmers to prioritise indicators which they assess as more reliable and science-based over those that are more related to cultural traditions. However, the importance of local knowledge lies not just in its scientific value, but also its role in enhancing the confidence with which farmers can relate the meteorological forecast to their local conditions and apply this local combined forecast effectively to agricultural decisions.

Putting the forecast into use

On a seasonal basis, the agricultural decisions that farmers can apply the forecast to fall broadly into two categories. First, largely irreversible decisions taken before or as the season starts, such as seed purchase and planting times, when seasonal forecasts are most useful, together with a significant number of local indicators. In central Tanzania, six of the 15 local indicators focused on the start of rains, second only to the number used to predict drought. Second, decisions made during the season, such as the timing of fertiliser application and pest control, which can be assisted by shorter-term forecasts. The examples cited have shown productivity increases in the range of 9–48 per cent, linked to decisions about crop variety, cultivation method and soil fertility management. Ensuring the usefulness of the forecast means providing it at the right time and in the right format. The CCAA-supported project in Benin (Ziervogel and Opere 2010) highlighted the need for seasonal forecasts at least a month in advance and at the right time of day. Farmers cultivating their land would only listen to the radio in the evening. Simple announcements without explanation or linkage to agricultural recommendations also limit a forecast's usefulness.

Developing an agricultural forecasting service

While the seasonal forecast acts as a suitable entry point to engage with farmers, there are clearly a variety of climate information products that need to be provided, including short-term (three-, seven- and ten-day forecasts), as well as monthly seasonal forecast updates. However, village development planning also needs more long-term information on five- or ten-year timescales to support decisions relating to investments that provide longer-term resilience benefits, such as setting up farmer groups and cooperatives, local reforestation and rehabilitation of degraded land areas and establishing irrigation systems. This increases community confidence that these investments are not maladaptive and plans can be used as inputs to local government planning and national adaptation strategy development processes, drawing down adaptation funding to the most vulnerable areas.

International support from climate funds and institutions

Where climate stations become local resource centres, rather than just measuring points, this also pays dividends. Farmers establishing rain gauges have access to training on their management and interpretation of their own data, as well as a link that can be used to access climate information, such as seasonal and short-term forecasts. This is clearly an option that can be expanded, bringing in agricultural advisory services (and other potential intermediaries) to blend agricultural recommendations with forecasts, thereby developing a 'farmer forecast' service that uses all relevant channels of communication, including local radio (and local languages broadcast services), local climate stations and agricultural offices, text messaging services, farmer groups and cooperatives. In some areas, early warning systems already exist that can be expanded to provide farmer forecasts. Farmers highlight soil fertility management and seed choice and supply as particularly important information areas to be linked to farmer forecasting services.

There is urgent need for a restructuring and strengthening of services to support agricultural adaptation, re-orienting the priority of global observing systems and national meteorology agencies away from a narrow focus on the needs of experts and policy-makers to those of farmers, herders and vulnerable communities. The World Meteorology Organisation's Global Framework for Climate Services Expert Panel Report highlights the importance of climate information tailored to the needs of users, confirming that 'there is a strong demand for climate services to deal with climate change and adaptation, particularly on the local level. There is a need for better ways of combining climate change projections with local climate data and knowledge.' As a result, Principle 2 states that 'the primary goal of the Framework will be to ensure greater availability of, access to, and use of climate services for all countries'. With an increasing number of successful projects focusing on the provision of specific climate services to farmers, the meteorology agencies, farmer organisations and groups, NGOs, donors and other key stakeholders involved need to share their experience, methodologies and best practice in the development of this comprehensive farmer forecasting service, supported by the Global Framework and the emerging system of climate change funding mechanisms.

Note

1 This has been found to correlate well with the movement of the ITCZ and so the start and occurrence of rains.

References

Banerjee, S.K., N. Chattopadhyay and H.P. Das, 2003: *Study of weather-based agricultural folklore of West Bengal*. India Meteorological Department.

Boer, Rizaldi, Kusnomo Tamkani and A.R. Subbiah, 2003: Communicating Climate Forecast to Farmers through Climate Field Schools: Indonesian Experience. www.climateadaptation.net/docs/papers/Rboer-Subbiah%20paper.pdf (accessed 22 September 2013).

Cheke, R.A. and J.A. Tratalos, 2007: Migration, Patchiness and Population Processes. *Bioscience* 57 (2): 145–154.

Ewbank, R., 2011: Review of the Innovations Fund Projects on Climate Change Adaptation 2008–10. http://unfccc.int/files/adaptation/application/pdf/christian_aid_review_of_the_innovation_fund_projects_on_climate_change_adaptation_2008-10_160811.pdf (accessed 22 September 2013).

Gurung, G. and Bhandari, D. 2008: An Integrated Approach to Climate Change Adaptation. *LIESA Magazine* 24 (4): 6–8. http://www.preventionweb.net/files/7880_ClimateChangeNepal.pdf (accessed 22 September 2013).

Helmuth, M., D. Zan Diarra, C. Vaughan and R. Cousin, 2010: *Increasing Food Security with Agro-meteorological Information: Mali's National Meteorological Service Helps Farmers Manage Climate Risk*, World Resources Report.

Kilongozi, N., Z. Kerenga and S. Leshongo, 2005: *Knowledge in Range Management and Forage Plants for Improving Livestock Productivity and Food Security in the Maasai and Barbaig Communities*, FAO.

Pat, A., P. Suarez and C. Gwata, 2005: Effects of seasonal climate forecasts and participatory workshops among subsistence farmers in Zimbabwe. *Proceedings of the National Academy of Sciences*, 102 (35): 12623–12628.

Practical Action Nepal, 2010a: *Impacts of Climate Change: Voices of the People – Based on Field Observations, Information and Interactions with the Communities in Nepal.* http://practicalaction.org/media/view/16890 (accessed 22 September 2013).

Practical Action Nepal, 2010b*: Temporal and Spatial Variability of Climate Change Over Nepal*. http://practicalaction.org/file/region_nepal/ClimateChange1976-2005.pdf (accessed 22 September 2013).

Ziervogel, G. and Opere, A. (eds), 2010: *Integrating Meteorological and Indigenous Knowledge-Based Seasonal Climate Forecasts for the Agricultural Sector*, Ottawa: IDRC.

9 Serious fun

Scaling up community-based adaptation through experiential learning

Pablo Suarez, Janot Mendler de Suarez, Bettina Koelle and Maxwell Boykoff

If adaptation to climate change involves people deciding to do things differently, then the challenge of scaling up community-based adaptation (CBA) is essentially how to create and communicate knowledge so it can be turned into action. As CBA practitioners, how can we accelerate knowledge-sharing dialogue and learning processes? How can we help subsistence farmers, humanitarian and development workers, government officials, donors and other key players to navigate the complex range of plausible climate risk management choices and outcomes? How can we meaningfully engage such diverse people and organisations so that they can access, understand, trust, generate, utilise and disseminate knowledge about adaptation?

This chapter addresses the challenge of scaling up CBA through the lens of participatory games and other *experiential learning* activities designed for engagement with climate-related problems – and possible solutions. These innovative climate communication strategies hold great potential to effectively translate complex and formal climate science-policy concepts in ways that resonate and 'stick' with everyday citizens, particularly those at the forefront of climate impacts. Ultimately, we argue that effective deployment of communications on climate change – in particular through experiential learning activities – may be as important as the governance architectures and low-carbon technological innovations to the success or failure of sustained engagement in dealing with contemporary climate change.

We first review the current understanding of some known communication obstacles, then examine some innovative approaches to overcome these challenges, with particular attention to game-based activities that enable people to inhabit complex systems where decisions have consequences. Our concluding discussion about desirable next steps to create an enabling environment for scaling up adaptation underscores the contribution of experiential learning activities to promote adaptive learning and knowledge-sharing dialogue in the adaptation process.

Understanding the communication challenges

The IPCC *Special Report on Managing the Risks of Extreme Events and Disasters to Advance Climate Change Adaptation* (IPCC 2011: 13) states that 'Measures that provide benefits under current climate and a range of future climate change scenarios', including early warning systems, are starting points for addressing projected trends in climate extremes. From short-term storm forecasts to long-term climate trends, organisations working at the community level have, as a result of scientific and technological advances, significant ability to anticipate threats to people at risk (e.g. Ghosh 2009). At the same time, the complexity and range of decisions that can be made both by people at risk and by the organisations serving them is rapidly expanding, owing to progress in technologies to obtain, process, communicate and use relevant information (Suarez 2009). From mobile phones and audiovisual equipment to participatory methodologies, we can produce and disseminate information and ideas useful to communities at an unprecedented scale. Yet, paradoxically, exacerbation of the 'climate problem' continues to outpace implementation of community-based adaptation measures to manage change and growing uncertainties. This does not mean to imply that effective implementation is a mere communication challenge; rather, we argue that climate communications comprise an often overlooked yet critical component of efforts to enhance multi-scale adaptation strategies over time. By extension, these considerations prompt an important question: what can be done to improve, promote and scale-up adaptive thinking and action on the ground?

Initial steps involve awareness-raising, and improving citizen understanding of the challenges ahead. Just as atmospheric scientists have identified patterns that shape the future of our global climate, social scientists have identified decision patterns that affect the perception, dissemination and utilisation of new ideas. For example: heuristics, or simplified procedures for decision support, are based on judgements about the likelihood of uncertain events (Tversky and Kahneman 1974). Heuristics prove acceptable most of the time, but may lead to undesirable choices in a variety of situations (Patt and Zeckhauser 2002).

Decision-makers tend to steer clear of engagement or involvement in issues and challenges where losses are inevitable. This has been called 'motivated avoidance' (Shepherd and Kay 2011). Additionally, people often evaluate a decision to commit an action (a commission, such as preparing for a disaster) more negatively than a decision to omit an action (an omission, such as ignoring warnings). Omission bias is a preference for greater losses arising through errors of omission over smaller losses associated with direct action (Spranca *et al.* 1991). Tetlock and Boettger (1994) demonstrated that accountability amplifies the status quo effect when change creates victims, even when implementing change could save many more lives. Their study found that lives lost from inaction were considered less valuable than lives lost from changing the status quo, with a ratio approaching nine to one in a scenario where subjects were accountable for modifying the existing state of affairs.

Given this built-in predisposition to weigh the risks of potential individual actions more heavily than the risks of inactions, it is no surprise that CBA is not growing fast enough. 'Mass media' – television, films, books, flyers, newspapers, magazines and radio – have provided spaces where public discourses have the potential to scale up from individuals to communities and societies (Boykoff 2008). Opportunities are also embedded in the power of the diverse and dynamic traditional/legacy mass media sources and institutions: the ability to serve as 'mediating' forces between science, policy and public communities; and the potential for directly and indirectly reaching unparalleled numbers of citizens-as-media-consumers.

Indeed, research has considered how media messaging can potentially influence attitudes and behaviours of public citizenry, empowering people to engage in or resist solutions for climate challenges (e.g. Boykoff 2011; Kahan *et al.* 2010). Moreover, recent analysis of media bias with respect to climate change denial suggests two factors which are likely to override such bias: direct personal experience with weather extremes, and the perception of shared knowledge about climate change (Painter 2011). Meanwhile, community-based organisations (CBOs) recognise that their grassroots communication efforts on climate adaptation cannot rely solely on the media, nor on formal primary and secondary education, to rapidly deploy the knowledge needed to scale up adaptation. These more conventional means of communication are seen to largely:

- depend on linear, unidirectional flows of information;
- view members of the public citizenry as passive, merely acquiring or recalling information;
- focus on the individual, rather than community locus, thereby offering limited, if any, interaction between individuals and groups who will ultimately need to work together on managing climate risks at the local level.

A different approach is needed to support learning at the pace and scale demanded by ongoing climate challenges. In other words, 'innovation' need not be limited to technological advancement; cultural 'innovations' can also help to address twenty-first century climate adaptation. Yet effective and sustained adaptation also requires a new praxis, i.e. 'reflection and action upon the world in order to transform it' (Freire 1974: 36).

Pilot projects abound. The question confronting the CBA community is how to scale up what works. We submit that experiential learning must be part of the answer.

Experiential learning to anticipate change

In a seminal work on experience as the source of learning and development, Kolb (1984: 38) defined experiential learning as 'a process whereby knowledge is created through transformation of experience'. This approach has been successfully applied to many disciplines and demonstrated to improve professional practice and the learning process (Kolb, 1984). It requires the resolution of conflicts between dialectically opposed modes of adaptation to the world: Kolb argues that learning requires abilities that are polar opposites, such as concrete experiencing of events at one end and abstract conceptualisation at the other, or reflective observation *and* active experimentation. In other words, if we carefully craft experiences that inspire opposing ways of dealing with the world, learning emerges naturally from the resolution of these conflicts.

CBA involves dynamic and dialectical elements that create tensions between differing valid ways of understanding climate-related issues: long term versus short term, individual versus collective, local versus national or agency versus structure. These opposing forces, including their nuanced interplay, are not easy to grasp through conventional, linear educational approaches. Trade-offs, feedbacks, non-linearities, delays and unanticipated 'side effects' are inherent in risk management decisions (Gonçalves 2008), and should be part of the learning experience of government officials and subsistence farmers alike. How does one devise a communication platform that can successfully convey the existence and relevance of system complexity?

We know that scaling up CBA requires a multi-stakeholder approach, with inherent obstacles in terms of communication and interactions across sectors and disciplines. Where complexity and access to information is a challenge, there is a need to establish innovative platforms to support community-based knowledge generation and sharing. As such, everyone can benefit from innovative learning approaches, especially when dealing with complex, dynamic and often vexing climate adaptation challenges.

A variety of processes are appropriate for facilitating experiential learning for adaptation to climate change: some of these approaches draw on the participatory learning and action (PLA) methodology (Pretty *et al.* 1995) and combine this with participatory action research (PAR) so as to increase resilience through learning (Kemmis and McTaggart 2000; Wadsworth 2001; Koelle and Oettle 2009). The experiential learning process can thus increase resilience by enhancing the capacity to anticipate future change of complex systems and to prepare for such possible change, informed by a collective perception of which future scenarios are most likely to materialise. Of course, the ability of communities to respond appropriately is determined by access to assets (Tschakert and Dietrich 2010).

A learning approach that is focused on detecting change and assessing the most appropriate responses to experienced and anticipated change fosters conditions in which members of affected communities are able to take charge of designing and implementing their own responses. This encourages optimism, resulting in greater empowerment and self-determination.

An experiential learning approach tested successfully in South Africa has utilised quarterly climate change preparedness workshops to create a platform where farmers share climate data, reflect upon observed weather patterns of the previous three months, critically interrogate seasonal forecasts and plan interventions, including innovative adaptation strategies. By following an action research approach, farmer-led learning has included on-farm experimentation in collaboration with scientists (Archer *et al.* 2008; Koelle and Oettle 2009).

Games for CBA: five cases

Albert Einstein once said that 'Games are the most elevated form of investigation' (McGonigal 2011). Well-designed games, like adaptation measures, involve decisions with consequences (Suarez *et al.* 2011; Mendler de Suarez *et al.* 2012). Participatory games are systems made up of interconnected parts which work together through a combination of rules, goals, narrative content, symbols and the delivery platform (whether it be a deck of cards, people's own bodies moving about a field or conference room, or a website) and can help people and organisations investigate a range of CBA choices and associated outcomes. Games can be designed so that simple rules can set into motion a gameplay experience filled with variety, novelty and surprise, from which meaning emerges – what Salen and Zimmerman (2004) describe as emergent complexity. Through games we can learn how systems work, and the game-based system rewards us as we learn (Macklin and Sharp 2012). Players inhabit, enliven and interpret these systems through play, and are readily compelled to learn how a game works for the sake of pleasure, discovery, competition and a whole host of other reasons that can be compressed into the notion of 'fun'. Games can train us to take a longer view, to practice complex thinking, and to examine multiple strategies aimed at solving planetary-scale problems (McGonigal 2011: 348).

This section shares case studies where participatory games were designed and facilitated to support rapid and effective communication of CBA concepts through experiential learning, involving members of vulnerable communities, scientists, government officials, donors, humanitarian and development workers and other stakeholders. These examples draw from gameplay experiences in Africa, the Americas, Asia and Europe, with CBA games addressing (1) forecast communication, (2) forecast use, (3) investments in climate risk management, (4) mosquito-borne diseases and (5) risk financing instruments.

Forecast (mis)communication

People continue to suffer and die due to entirely predictable extreme events. Notwithstanding the ambition to take humanitarian action *before* a disaster or health emergency happens (IFRC 2009), forecasts regularly go unused. Regrettably, conventional approaches to training workshops on science-based forecasts for risk management (such as PowerPoint presentations from decision-averse scientists to decision-oriented practitioners) tend to lead to utter confusion, or simple boredom.

The Red Cross/Red Crescent Climate Centre joined forces with game designers from the Parsons School for Design to co-create experiential learning processes for linking early warning to early action. The occasion was a four-day workshop held in northern Senegal in December 2009. It convened about 40 people who would not normally talk with each other: forecasters, Red Cross personnel and members of a vulnerable community. The event was carefully designed using participatory games to create an atmosphere of collaboration without hierarchies, in which every participant had something to learn and something to teach.

When conceiving ways to link knowledge with decisions, many stakeholders assume messages will travel in a reliable and timely manner from the source (such as experts or government authorities) to the receiver (e.g. communities at risk), yet we know that the information content will likely be distorted as the message travels through various channels. 'Spreading the word' is a game on communication gone awry, a serious yet fun learning experience that irrefutably demonstrates to forecasters and users the risks inherent in communicating information about climate risks. Adapted from a classic party game about the degradation of information over time and distance known as 'telephone' or 'Chinese whispers', the facilitator writes a sentence involving CBA on a piece of paper, and invites the first person from each row to read and memorise the short text (i.e. 'the message' that each team must communicate down the row). Then each person in the row conveys the message only to their neighbour, and only once. When the last person receives the message, she writes it as heard (the 'end message'), and brings the text to the facilitator, who notes the order in which each team completes the task. The winning team is the first to communicate the message without any consequential distortions (Figure 9.1).

When players learn the rules before gameplay, they are told how crucial it is that the message not be altered, and to ensure there is no distortion. Teams immediately understand the tension that emerges between *speed* in communicating the message (which determines which team wins) and potential distortion along the communication chain (which can lead to a team's elimination). Typically, at least some of the teams' end message is radically different from the original message – resulting in hilarious outcomes – illustrating how easily crucial communications can break down.

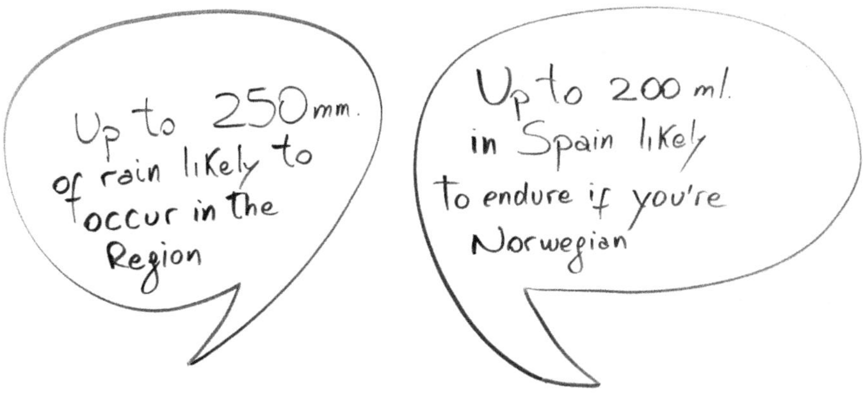

Figure 9.1 Distortion of science-based forecasts was vividly experienced by scientists, humanitarian workers and villagers playing the game 'spreading the word'.

This game was played with facing teams sitting on both sides of a long table. A senior member of Senegal's National Meteorological Agency was asked to write the message (in French), roughly translated as 'There is an 80% chance of at least 250 millimetres of rainfall during the next 72 hours in the north of the Saint-Louis Province' (a forecast of extreme rains, worth acting upon for flood preparedness). The message travelled down the rows and the last person in each team shared the message with all participants. The fastest team turned the message into an enigmatic sentence involving vegetables, drawing considerable laughter. The slower team perpetrated only one minor modification: instead of '250 millimetres' the end message was '50 millimetres'. All teams had failed to communicate the original message and thus there was no winning team.

The significance of these distortions emerged among players, first as an entertaining recognition of which team had done better, and then the 'Aha!' moment sunk in: these participants were actual producers and users of climate predictions, and each player was convinced of having relayed the message correctly. Yet, even for the best-performing team, the distortion meant the difference between catastrophic rain (original message) and a normal event, not worth acting upon (end message). Rich conversation spontaneously ensued about how experts, Red Cross and community members should work together to anticipate the risk of miscommunication, and design early warning systems that are both robust and quick in delivering information. Experiential learning transformed collective understanding of practical challenges, abilities and limitations in communicating forecasts and, by extension, CBA.

Forecast use

Producers and users of forecasts have very different languages, perspectives and priorities, and are not used to jointly examining the issue of whether or not to act based on a given forecast expressed in terms of probabilities. As a result, users often cannot understand the meaning and implications of experts' statements about likely future conditions, and scientists cannot understand why their forecasts are not used. To open a dialogue process whereby forecasters and users can understand each other's way of understanding the role of forecasts for informing action, a game can establish a playful atmosphere as a safe space for expressing disagreement and finding or creating common ground for future, long-term collaboration.

'Early warning, early action' is a decision-making card game about climate risk management at different timescales, designed for participants to explore different perspectives on how to link forecasts with action.[1] It was designed for the Senegal workshop described in the previous case study. The learning experience takes about 45–90 minutes, depending on how much time is dedicated to discussion.

People form groups of 5–8 players around a table, with forecasters and users in each group. In each round, one of the players serves as 'Decider', and all other players are 'advisors', receiving four cards from a deck of illustrated *action*

Figure 9.2 Members of a vulnerable Senegalese community play the 'early warning, early action' card game with forecasters and Red Cross staff.

cards. Each action card has an idea for a task that could be carried out before a disaster materialises, such as 'evacuate', 'pre-position tents for shelter' or 'wait and see'. Then three forecasts are revealed, showing expected future conditions with different lead times – and different levels of complexity, such as 'The village upstream is flooded, and the peak flood may arrive at your village in 12 hours', or 'Up to 35 mm accumulated rainfall expected with 80% probability in this region for the coming ten days'.

Advisors must recommend a plan for disaster preparedness. For each forecast, each advisor chooses one action card from their hand (or writes their own action card if they drew a blank card), placing it face down. The Decider chooses from all the recommended action cards the one she thinks most appropriate for each forecast. Advisors whose cards are chosen receive one point for each pre-written action card and two points for each player-generated action card selected. Advisors may challenge the Decider's choice, generating discussion as the challenger advocates for her action, the Decider defends her decision and other players vote to uphold or overturn the Decider's choice. Forecasters, humanitarian workers and community members engaged fully in selecting among multiple plausible forecast-based actions, discussing their merits and risks – including what can go wrong either in terms of 'acting in vain' or 'failing to act'. Differences of opinion led to rich conversations about the links between what science can say and what people are willing and able to do given vulnerabilities, capacities, cultural norms and many other aspects shaping whether early warnings trigger early action.

The Senegal Red Cross then took workshop participants to the island village of Doun Baba Dieye in the mouth of the Senegal River. The game was played in groups with a majority of local women. Players created over 300 new action cards, generating ideas for disaster preparedness at the community level. Importantly, people felt comfortable exploring and expressing their views, regardless of scientific credentials or levels of seniority.

Investments in climate risk management

The Rockefeller Foundation has embraced the concept of 'resilience' as one of the pillars of its work. Its website states that 'building resilience goes beyond the concept of adaptation, and refers to the capacity over time of a system, organisation, community, or individual to create and implement multiple adaptive actions'. 'Dissolving disasters: A resilience game where donors walk the talk', was commissioned as part of a series of experiential learning workshops about the concept of resilience and its implications for the work of the Rockefeller Foundation.[2]

Assuming the role of subsistence farmers organised in community teams, players win and lose beans depending on simulated rains – and on their decisions about crop selection. Planting investment decisions are executed by walking to the sector of the room designated for one of three choices:

- *Flood protection*: invest in flood-resistant crops (e.g. rice).
- *Drought protection*: invest in drought-resistant crops (e.g. cassava).
- *No disaster protection*: invest in crops that perform best if no drought or flood occurs (e.g. maize).

Drought and flood protection require an upfront payment of one bean. No protection has no upfront cost, but if there is an extreme event then unprotected farmers must pay four beans to recover. If there is no extreme event, or if a flood or drought occurs but the farmer is protected against it, she gains two beans that round. A die is rolled to represent the probability of disaster occurrence: if a six is rolled, there's flooding; a one is drought; all other rolls mean normal rains. Farmers who can't afford to pay after a disaster must leave the village, and the game.

A few participants take the role of a donor organisation trying to support a community, with limited resources and three distinct choices:

- *Disaster response*: after a disaster, give beans to needy farmers.
- *Disaster preparedness*: before the rains, give farmers incentives for flood or drought protection.
- *Growth*: help farmers to accumulate beans if no disaster occurs.

After a few turns, a climate change variation is introduced: the die is replaced by a truncated cone. The occurrence of a flood, drought or normal year is determined by the way the object lands after spinning in the air. Players find themselves at a loss estimating the new probabilities of rainfall risks, and often competing theories emerge about most likely outcomes.

Figure 9.3 The probability of an extreme event in the game 'dissolving disasters' is first portrayed by a roll of the die, and then by a less-understandable object to represent climate change.

From the outset, players realised the need to discuss individual and collective strategies within their team. Tensions emerged between donor and farmer perspectives on risk taking versus risk reduction, which had to be negotiated through thoughtful argumentation about possible consequences of different decision options. Because of the counterintuitive probability of occurrence of different outcomes under climate change, many players found themselves wiped out of the game after only a few rounds. Players lived through the changing probabilities, and experienced the consequences of not knowing what to expect. This powerfully conveyed to participants the need to anticipate and manage changing risks when designing and implementing humanitarian and development work.

After about an hour of gameplay wearing someone else's shoes and walking the talk, players think and feel differently about resilience. As mentioned by a Foundation staff member during game debriefing: 'It was an "aha!" moment for me … *I was a farmer*.' The game ends after the sixth round, with winners and losers: the player with the most beans is the individual winner, and the community with most beans is the winning team. The winning donor is the one that supported the village that loses the least number of farmers. These incentives create trade-offs between collaboration and competition, and mimic complex

Figure 9.4 Malawian subsistence farmers learn the basics of index-based micro-insurance in a game using coupons, a roll of the die and real money.

feedbacks and thresholds that trigger rich discussions delving into resilience and adaptation concepts.

This participatory game was deliberately designed with system flexibility in order to enable people and organisations to explore modifications to game rules and narratives aimed at better capturing aspects of specific relationships between context, decisions and consequences. It has been successfully adapted for numerous other climate-related events on four continents. Tailor-made versions were run, for example, for an Oxfam staff retreat in Oxford, a symposium on climate impacts for the government of the Province of Buenos Aires, a UNFCCC workshop on disaster risk reduction in Lima and the Communications Day event during the 2011 UNFCCC conference of the parties in Durban (Reuters 2011).

Mosquito-borne diseases in a changing climate

The 2011 UNFCCC conference in Durban offered a new, seriously fun way to learn about health risks and climate change: a new participatory game developed for the RC/RC Climate Centre by a team of graduate students and faculty at Yale University and Parsons The New School for Design. Half the players are dengue-carrying mosquitoes attempting to bite humans and lay eggs to reproduce. The opposing team are humans attempting to protect themselves from being bitten and to clean up mosquito breeding grounds – all while a changing climate affects their chances of success (Tran *et al.* 2011) (Figure 9.2b).

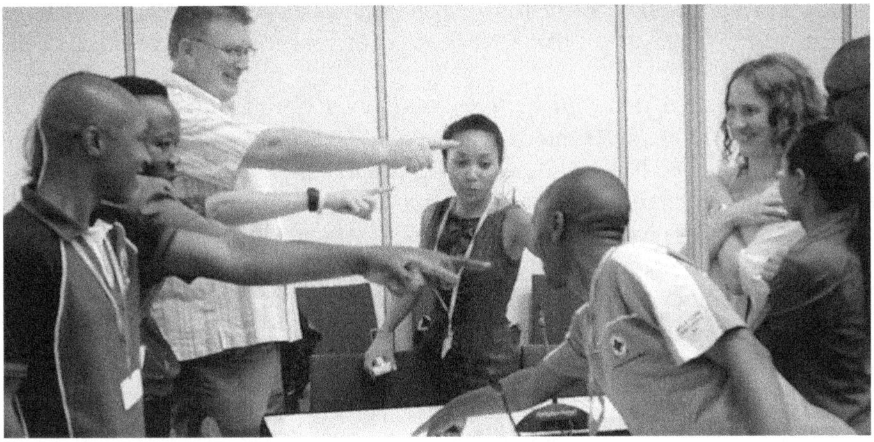

Figure 9.5 Parliamentarians, climate experts and Red Cross youths play 'humans vs. mosquitoes', a rock–paper–scissors-like climate game.

In four sessions, the game 'humans versus mosquitoes' engaged about 40 participants, including Ugandan and Argentinean parliamentarians, Red Cross youth from Africa and the Americas and senior IFRC managers, who are now working on plans to scale-up use of the game in schools and with Red Cross volunteers.

Importantly, the rules of this quick and intense game capture the dynamics of climate and health risk management decisions and associated consequences. It is deliberately designed to be playable in vulnerable communities, and adaptable to other climate-sensitive vector-borne health threats.

Financial instruments for climate risk management

The World Food Programme commissioned a game to help people in rural communities acquire a working understanding of index-based micro-insurance (a risk-sharing instrument tailored to the needs of the poor, which might help vulnerable farmers deal with extreme events). Altogether 168 subsistence farmers from two communities in Ethiopia and two in Malawi (many illiterate and even innumerate) made decisions in a simulated drought insurance market linked to loans for agricultural inputs. Gameplay employed coupons, a die and real money. Players had to examine the trade-offs between the costly option of buying insurance versus the riskier option of hoping for good rains. Farmers in all play-sites found the game readily understandable and engaging – and quite entertaining. According to one of the most experienced Agricultural Extension Officers involved in marketing a cognate micro-insurance pilot in Malawi, 'the game makes it much easier for us to explain to farmers how insurance really works'.

Analysis of the post-game survey revealed that the game was at least as good as a conventional lecture approach in conveying most of the key insurance concepts – and significantly better at one key dimension: trust (Patt *et al.* 2010). 'Diving into the regional insurance pool' is a similar game successfully played with delegates to the UNFCCC conference in December 2010 in Cancún.[3]

Oxfam America, IRI and partners went even further in a project called HARITA, which integrates micro-insurance into a broader climate risk management package: game-based tools supported collaboration with Ethiopian farmers, eliciting their preferences and facilitating a dialogue process that led to several key innovations in product design, including 'Insurance for Work' (farmers who cannot afford the premium can pay for it with monetised labour), and linking risk sharing with risk reduction (monetised labour directly contributes to reducing local vulnerability to future extreme events through measures such as irrigation and soil conservation). Another game, called 'FAIR', was designed to help donors and other stakeholders understand how HARITA supports farmer-led CBA. Patt *et al.* (2009) describe four game-based research endeavours for micro-insurance in Africa, Asia and South America.

The way forward

These trends in new learning approaches are emerging in the context of a wider and fundamental set of questions involving how communication strategies on climate change can be most effective in meeting people where they are, while inspiring and enabling new considerations, capacities and behavioural changes. This chapter has sought to articulate the complexities involved in such endeavours, and has outlined both limitations and opportunities for more effective communication and connected actions that scale up from community levels to a wider scope of adaptation.

The case studies illustrate how recent developments in participatory methods are demonstrating enormous potential for improving the co-production and sharing of information and insights for CBA, by reinforcing both direct personal experience and the perception of shared knowledge. *Experiential learning* is an approach that can enable those managing climate risks to address the complexity of the relevant system through activities that link decisions with consequences, combine collaboration and competition, reveal differences and commonalities in understanding problems and solutions, and generate collective intelligence and new knowledge.

Experiential learning can bring a range of different stakeholders together and encourage improved communication and understanding while the 'learning partners' discover joint interests and accept each others' differences. It is these partnerships that support synergistic approaches to address the complexity in CBA. While experiential learning approaches can offer effective tools to support adaptation, we caution practitioners and community members alike: in order to moderate meaningful experiential learning processes, skilled facilitators are necessary so as to ensure that all participants in the learning process feel they can mindfully and openly share, learn and explore in a safe space. Issues of power, gender and hierarchy are often very challenging – especially when operating under ongoing social and environmental change and high levels of uncertainty regarding future framing conditions. These factors need to be taken into account when designing and implementing experiential learning processes. It is important to recognise that participation in learning processes is time consuming for participants and exposes them to potentially uncomfortable or challenging situations that they are not able to anticipate and might only be able to steer to a limited extent. Participation in these learning experiences thus requires some risk taking on the part of the participants, and unless this experience and the learning is perceived to be valuable, the participants are unlikely to engage in similar processes in the future.

The experiences and insights of associated approaches and methodologies (psychology, participatory development, disaster risk reduction, etc.) provide valuable resources for innovative learning that will allow us to effectively and creatively implement CBA. However, whatever methodologies and tools are applied, success will depend on skilled and capable facilitators who are sensitive to the cultural and gender dimensions of the community. Furthermore, it is unlikely that stand-alone and one-off interventions will result in profound change: optimally, experiential learning processes should be embedded in ongoing processes in support of CBA.

Acknowledgements

We are grateful to our colleagues in the humanitarian, development and academic sector who supported the design and facilitation of experiential learning activities for climate risk management, especially from IFRC, IRI, Oxfam America, Parsons The New School for Design, Red Cross/Red Crescent Climate Centre, Rockefeller Foundation, UNFCCC Secretariat, WFP and the World Bank. We also thank those who helped create the enabling environment for integrating participatory games into adaptation initiatives, especially Colleen Macklin, Cristina Rumbaitis del Río, David Sattherthwaite, Jane Clark, Kanta Kumari, Maarten van Aalst, Madeleen Helmer, Margaret Arnold, Miwa Kato, Mónica Casanovas, Richard Klein, Saleemul Huq, Sirkku Juhola, Ulrich Hess and Youcef Ait-Chellouche.

This chapter was developed as part of a research grant from the Climate and Development Knowledge Network (CDKN Action Lab Innovation Fund). As such, it is an output from a project funded by the UK Department for International Development (DFID) for the benefit of developing countries. However, the views expressed and information contained in it are not necessarily those of or endorsed by the DFID, which can accept no responsibility for such views or information or for any reliance placed on them.

Notes

1 See a four-minute video of the game at www.youtube.com/watch?v=Mpj_EbKdwEo, and the full set of rules at http://petlab.parsons.edu/redCrossSite/gamesBTS.html.
2 A seven-minute video about the game is online at www.youtube.com/watch?v =wKHiTV9TaAQ.
3 A six-minute film showing delegates at play is online at http://vimeo.com/27571755.

References

Archer, E.R.M., N.M. Oettlé, R. Louw and M.A. Tadross (2008). '"Farming on the edge" in arid western South Africa: climate change and agriculture in marginal environments'. *Geography* 93(2): 98–107.

Boykoff, Maxwell (2008). 'Flight semantic drift!? Mass media coverage of anthropogenic climate change'. In Michael K. Goodman, Maxwell T. Boykoff and Kyle T. Evered (eds), *Contentious geographies: environmental knowledge, meaning, scale*. Hampshire and Burlington, VT: Ashgate Publishing Ltd.

Boykoff, Maxwell T. (2011). *Who speaks for the climate?* Cambridge: Cambridge University Press.

Freire, P. (1974). *Pedagogy of the oppressed* (M.B. Ramos, trans.). New York: Seabury Press.

Ghosh, Aditya (2009). 'The sunshine industry'. *Hindustan Times*, 5 April: 14.

Gonçalves, P. (2008). 'System dynamics modeling of humanitarian relief operations'. MIT Sloan School Working Paper 4704-08, http://ssrn.com/abstract=1139817, accessed 2 January 2012.

IFRC (International Federation of Red Cross and Red Crescent Societies) (2008). 'Early warning > early action'. Online at: www.ifrc.org/Docs/pubs/helpnow/early-warning-early-action.pdf (accessed 10 August 2010).

IFRC (International Federation of Red Cross and Red Crescent Societies) (2009). *World Disasters Report 2009: Focus on Early Warning Early Action*. Online at: www.ifrc.org/publicat/wdr2009/summaries.asp (accessed 19 September 2013)

IPCC (Intergovernmental Panel on Climate Change) (2011). *Special report on managing the risks of extreme events and disasters to advance climate change adaptation. Summary for policy makers*. Online at www.ipcc-wg2.gov/SREX (accessed 25 July 2013).

Kahan, Dan M., H. Jenkins-Smith and D. Braman (2010). 'Cultural cognition of scientific consensus'. *Journal of Risk Research* 9(2): 1–28.

Kemmis, S. and R. McTaggart (eds) (2000). *Participatory action research: handbook of qualitative research*. London: Sage.

Koelle, B. and N. Oettle (2009). 'Adapting with enthusiasm: climate change adaptation in the context of participatory action research'. Paper presented at the Sustainable Land Use Conference, Windhoek, Namibia, Ministry of Environment and Tourism, Namibia.

Kolb, D.A. (1984). *Experiential learning: Experience as the source of learning and development*. New Jersey: Prentice Hall.

Macklin, Colleen and John Sharp (2012). 'Freakin' hard: game curricula about game design, issues and technology'. In Constance Steinkeuhler, Kurt Squire and Sasha Barab (eds), *Games, learning and society: learning and meaning in the digital age*. New York: Cambridge University Press.

Mendler de Suarez, J., P. Suarez, C. Bachofen, N. Fortugno, J. Goentzel, P. Gonçalves, N. Grist, C. Macklin, K. Pfeifer, S. Schweizer, M. Van Aalst and H. Virji (2012).

Games for a New Climate: Experiencing the Complexity of Future Risks. Pardee Center Task Force Report. The Frederick S. Pardee Center for the Study of the Longer-Range Future, Boston University. Online at: www.bu.edu/pardee/publications-library/2012-archive-2/games-climate-task-force (accessed 19 September 2013).

Painter, J. (2011). *Poles apart: the international reporting of climate skepticism*. Cambridge: Reuters Institute for the Study of Journalism.

Patt, A.G. and R. Zeckhauser (2002). Behavioral perceptions and policies toward the environment'. In R. Gowda and J. Fox (eds), *Judgments, decisions, and public policy*. Cambridge: Cambridge University Press.

Patt, A.G., N. Peterson, M. Carter, M. Velez, U. Hess and P. Suarez (2009). 'Making index insurance attractive to farmers'. *Mitigation and Adaptation Strategies for Global Change* 14: 737–757.

Patt, A.G., P. Suarez and U. Hess (2010). 'How much do smallholder farmers understand and want insurance? Evidence from Africa'. *Global Environmental Change* 20 (1): 153–161.

Pretty, J., I. Guijt and I. Scoones (1995). *A trainer's guide for participatory learning and action*. London: IIED.

Reuters (2011). 'Games wake people up to climate change', http://uk.reuters.com/article/2011/12/02/us-climate-games-idUKTRE7B10HY20111202, accessed 2 January 2012.

Shepherd, Stephen and Aaron C. Kay (2011). 'On the perpetuation of ignorance: system dependence, system justification, and the motivated avoidance of sociopolitical information'. *Journal of Personality and Social Psychology* 102 (2): 264–280.

Spranca, M., E. Minsk and J. Baron (1991). 'Omission and commission in judgment and choice'. *Journal of Experimental Social Psychology* 27 (1): 76–105.

Suarez, P. (2009). *Linking climate knowledge and decisions: humanitarian challenges*. Boston, MA: Boston University Frederick S. Pardee Center for the Study of the Longer-Range Future.

Suarez, P., J. Benn and C. Macklin (2011). Putting Vulnerable People at the Center of Communication for Adaptation: The Case for Knowledge Sharing through Participatory Games and Video Tools. World Resources Report 2011 – Expert Perspectives. Online at: www.worldresourcesreport.org/responses/putting-vulnerable-people-center-communication-adaptation-case-knowledge-sharing-through-p (accessed 19 September 2013).

Tetlock, P.E. and R. Boettger (1994). 'Accountability amplifies the status-quo effect when change creates victims'. *Journal of Behavioral Decision Making* 7 (1): 1–23.

Tran, L., C. Ewing, M.F. Dutta, B. Norskov, E. Labay, S. Colantonio, K. Shrestha, L. Graham and V. Lamers (2011). 'Humans versus mosquitoes', http://humansvsmosquitoes.com, accessed 2 January 2012.

Tschakert, P. and K.A. Dietrich (2010). 'Anticipatory learning for climate change adaptation and resilience'. *Ecology and Society* 15 (2): n.p.

Tversky, A. and D. Kahneman (1974). 'Judgment under uncertainty: heuristics and biases'. *Science* 211: 1124–1131.

Wadsworth, Y. (ed.) (2001). *The mirror, the magnifying glass, the compass and the map: facilitating participatory action research. Handbook of action research – participative inquiry and practice*. London: Sage.

Part III

Scaling up in action

10 Scaling up community-based adaptation to protect health from climate change

Joy Shumake-Guillemot, Kristie L. Ebi, Iqbal Kabir, Thuan Nguyen and Mazan Malkawi

Communities in developing countries and least developed countries (LDCs) are recognised as the most vulnerable and at risk for a range of negative health impacts associated with climate change (Confalonieri *et al.* 2007). Although vulnerability to health impacts of climate change exists in all countries, these communities already experience higher exposures to health risks associated with poverty, endemic infectious diseases, increased urbanisation and inadequate water and sanitation infrastructure. They are already burdened with the world's most prominent climate-sensitive health challenges, such as malaria, diarrhoea, and undernutrition (Confalonieri *et al.* 2007), and frequently have limited access to health care and preventive services. These challenges are projected to worsen with climate change, assuming limited adaptation and mitigation. In addition, many resources essential to health, such as adequate water and food security for these communities, are also threatened by changing climatic conditions. The World Health Organization, other organisations and scientific experts have made clear that adapting to current and future health risks posed by climate change through public health policy and interventions should now be a health sector-wide priority critical for future human development (Costello *et al.* 2009; WHO 2008). Therefore, the protection of vulnerable communities in developing countries should be central to efforts for health and climate change adaptation.

To prevent avoidable health risks exacerbated by climate change at the community level, increased attention, innovation and adequate resources are urgently needed to engage and assist vulnerable communities worldwide to scale up a range of public health, health care, disaster and climate risk management interventions. It is already recognised that many of the most effective strategies are ones that engage and empower local actors and use local information about a population's vulnerabilities and the risks to which they are exposed (Ebi and Semenza 2008). Thus, the use of approaches that both identify local needs and empower community actors to manage their own health risks will be essential to improving the overall reach and effectiveness of the health sector in any country. A range of additional actions will also be needed by national and international institutions and organisations. So how can global and national health and development actors best engage with and support communities to prepare for the future?

This chapter explores how the expanded use of community-based adaptation (CBA) approaches can help the health community be more effective at supporting communities under conditions of a changing climate. CBA approaches are based on the premise that local communities are equipped with existing skills, experience, local knowledge and social networks to implement locally appropriate activities that can increase resilience and reduce vulnerability to a range of factors, including climate change (Dodman and Mitlin 2011). The health sector is well placed to utilise CBA approaches because community health services are the foundation for protecting the health of the most vulnerable communities, and local practices for well-being are considered vital in many developing countries. The scale-up of CBA as part of climate risk management can help ensure that interventions designed, funded and managed by national or subnational health authorities adequately respond to local risks and community needs. This chapter describes three case studies from Vietnam, Bangladesh and Jordan, and highlights five common issues important for scaling up CBA to empower communities for health protection from climate change.

Background: reaching communities for climate adaptation in the health sector

The health sector is already working to protect populations from climate-sensitive health risks, such as undernutrition, vector- and water-borne diseases and injury or ill health during extreme weather events. Community health and primary health care services are the cornerstones of a health system in any country, with established mechanisms to reach, monitor, communicate with and protect individuals and communities. In order for community health systems to respond to climate change risks, current actions will need to be fortified and enhanced to reduce vulnerability and, at the same time, become prepared for possible additional risks, such as heightened disease transmission or health hazards from extreme weather events. To do so, active collaboration with other sectors and a range of additional financial, human and technical resources will be required for the health sector in a country to adapt to climate change.

Climate change adaptation initiatives in many countries are now beginning to identify climate risks to health and propose adjustments in health programming. These interventions span a wide range of activities, from adjustments to vector-borne disease control strategies to community preparedness for extreme weather events, owing to the locally complex nature of both climate change impacts and public health management. The adaptation measures that will be needed and appropriate in a specific area will be determined by the particular local and regional conditions of vulnerability, including the available coping mechanisms. Thus, because climate change impacts occur and need to be managed at the local level, the 'scaling up' of community-based efforts will be a significant aspect of scaling up overall climate change adaptation efforts in a country, particularly in the health sector.

The need to increase the impact and coverage of health promotion, prevention and services is a common challenge in public health. The traditional concept of 'scaling up' used in the health context often refers to 'deliberate efforts to increase the impact of health service innovations successfully tested in pilot or experimental projects so as to benefit more people and to foster policy and programme development on a lasting basis' (ExpandNet 2011; Syed *et al.* 2008; Hanson *et al.* 2003; WHO 2009). This is done by increasing the coverage of a proven intervention to a greater proportion of a population or diversifying the range of health services offered to address a health problem. Significant experience exists in how to scale-up community-level health interventions (ExpandNet 2011; Syed *et al.* 2008; Hanson *et al.* 2003; WHO 2009), which can be useful to scaling up CBA to climate change.

This chapter focuses on how to scale up CBA for health, meaning *to increase the use and effectiveness of CBA approaches that encourage or enhance participation and community-centred decisions about health, as well as provide and sustain actions that protect health at the community level.*

To address climate risks to health at the community level, navigation between national and community priorities will be required. The following sections describe the common interests and roles that actors at the community and national levels often play, followed by case studies of how to bring these actors together to keep community interests at the centre of development and risk management decisions.

Role of communities in climate adaptation for health protection

Ultimately the health consequences of climate change at the community level will depend upon pre-existing health status, the type of health risks to which communities are exposed and how well prepared a community is to cope with rapid and/or long-term changes in the occurrence of extreme weather events and diseases (Ebi *et al.* 2011). In many cases, climate change will act as a multiplier of existing health risks and exacerbate the magnitude and distribution of health risks already managed by communities. However, new threats can also emerge as the risk landscape changes, such as the introduction of vector-borne disease in previously unaffected areas or a change in patterns of existing health risks, like the frequency of extreme temperature events or seasonality of disease transmission.

This localised nature of health risks elevates the importance of involving communities to recognise and monitor local health risks, communicate risks, provide services and social support, propose and implement locally appropriate solutions and mobilise communities to maintain good health.

There are a wide range of actors and stakeholders involved in the delivery of community health services, including local authorities, non-governmental organisations (NGOs), community-based organisations (CBOs), private and public health service providers, the media, leaders and educators. Health care service providers can be from governmental, non-government, religious or private sector organisations. Community-based health workers recognised as health auxiliaries, health volunteers, health promoters, family welfare educators and village health workers serve a particularly important role in delivering preventive care, monitoring community health status and risks, identifying vulnerable and at-risk patients and, at times, provision of basic curative health services. Community-based actors, including individuals, community, state and national agencies, religious organisations, NGOs and others, can play a range of roles in adaptation, to reduce exposures, prevent disease onset and reduce negative health outcomes (Hanson *et al.* 2003).

It has been noted for decades that with appropriate support community health programmes can be an effective way to deliver critical primary health services at the local level (Haines *et al.* 2007), enhance health education and build social networks and community coping capacity to deal with emergencies and disasters (Keim 2008). Community-based programmes are particularly critical in low-income contexts where communities often have limited access to effective health systems and emergency services. Enabling communities to better manage their existing vulnerabilities and health risks is key to helping them manage changing health risks due to climate change.

Role of national-level actors for climate change adaptation and health policy

The magnitude of potential climate impacts on health (Confalonieri *et al.* 2007) are now being taken seriously by national governments and health authorities (Neira *et al.* 2008). Many countries across the world are preparing Health Sector Climate Adaptation assessments and national strategies, either as independent sector-specific exercises, or part of larger multi-sectoral efforts such as National Communications to the UNFCCC, National Adaptation Programmes of Action (NAPAs), and National Adaptation Plans (NAPs) in LDCs.

National-level actors play important roles in policy-making, agenda setting, strategic planning, research, resource mobilisation and allocation, legislation and regulation. These actors include government ministries and institutes, bilateral and multilateral donors and academic institutions that engage in capacity development and institutional strengthening at national and subnational-levels, coordinate across health and other sectors and generate, use and translate science into policy (e.g. epidemiological models, climate models, economic analysis).

National strategies that account for and address systemic, longer-term, cross-border and broader-scale multi-sectoral risks are necessary to obtain an aggregate and holistic picture of the range of vulnerable communities, districts, and populations in a country, and to estimate and identify future health risks. The adaptive measures included in national or subnational strategies can range from changes in regulations, to multi-sectoral collaborations that ensure key health determinants like food and water security, to actions that improve core health system functioning. National strategies can serve to strengthen health systems and better manage current climate-related health risks. Ultimately, however, top-down interventions must be implemented and translated in ways that are acceptable, understandable and respond to communities' needs and values. Thus, identifying and strengthening the mechanisms and partnerships that can translate and transmit information across scales from national to community level are essential to bring together efforts and information that can build community and health sector institutional capacity to address climate risks. The following case studies demonstrate why this is important and how it can be done.

Case studies of CBA for health

CBA approaches are highly compatible with and often similar to traditional community health approaches that manage climate-sensitive health risks. However, in the context of climate change these common approaches may need to reach beyond traditional partnerships and employ additional risk management practices. The following case studies from Vietnam, Bangladesh and Jordan illustrate the application of CBA approaches to prevention of health risks exacerbated by climate change, and highlight lessons for involving and empowering communities. These current health adaptation projects help inform how to increase the scope and reach of CBA approaches that protect health.

Vietnam: NGO and government collaboration for dengue control

Dengue fever, transmitted primarily by the *Aedes* mosquito, is a climate-sensitive disease already taking a significant toll on urban communities in Southeast Asia. Currently, there are no operational early warning systems for dengue outbreaks that can trigger early community action to identify cases and halt transmission. Therefore, it is essential to have effective community surveillance to identify early cases when outbreaks occur. In Southeast Asia, warmer and wetter conditions associated with climate change may alter the seasonality of dengue fever and increase transmission risks in certain areas (Benitez 2009). In Vietnam, the Ministry of Health and Vietnamese National Red Cross (VNRC) have observed an increase in the severity of dengue outbreaks in association with the severity of flooding.[1] The Red Cross/Red Crescent Climate Centre, through the International Federation of Red Cross and Red Crescent Societies and the VNRC, are working in partnership with the Ministry of Health of Vietnam to understand and communicate these changing risks and engage institutions and communities to effectively prevent health problems.[2]

To inform and evaluate community interventions the VNRC carried out a community-based survey in Ho Chi Minh city and Tian Giang about 'risk perceptions of dengue and climate change' (Red Cross Climate Centre 2010). These communities have high disease burdens and frequent flooding that result in pooling of water, which provides vector breeding sites. They reported that 48 per cent of community members surveyed believe that climate change will increase health problems and they are concerned about how to protect themselves (Nerlander *et al.* 2011). Survey findings helped inform the design of health communication materials. They concluded that main messages should continue to focus on evidence-based methods of dengue prevention, but that risk factors for changing patterns of disease, of which climate change is one, should also be discussed. Communicating uncertainty about how climate change will affect the intensity of transmission, or the timing and location of dengue, in the context of other local-scale risk factors remains a challenge.

In order to enhance health sector readiness to deal with surprises that climate change may present, this project also aimed to strengthen the existing collaboration between the VNRC and the Ministry of Health at national and provincial/district levels. They addressed this need by establishing three tiers of steering committees (Figure 10.1) at national, district and community levels to bring together the VNRC and Vietnam health authorities. Communication and reporting occurs internally to each institution, as well as between the levels of steering committees. This institutional arrangement has been a key mechanism to strengthening the partnership for dengue control at national and community levels. It enhanced the sharing of information on the distribution of dengue cases in Vietnam and coordination of response efforts to ensure that early detection is quickly translated into early responses to control local transmission and ensure access to care for those affected. The presence of VNRC community-based volunteers helps ensure that dengue prevention activities are ongoing while the response time to provide treatment in case of an outbreak is shortened. The community presence of VNRC volunteers coupled with collaboration with regional and national levels of the Ministry of Health, can improve the performance of dengue control programmes by strengthening the coming together of top-down and bottom-up actions.

Local NGOs such as the Red Cross often have high levels of community trust and access and are well placed to capture local realities, such as in Vietnam.

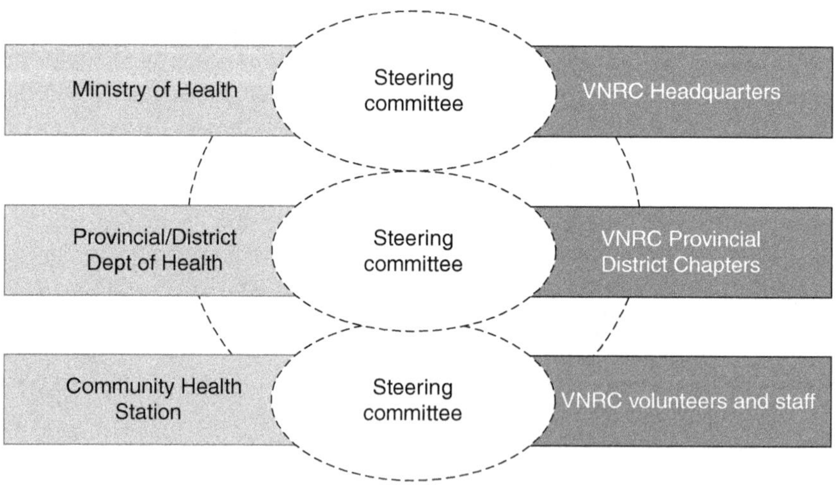

Figure 10.1 Vietnam: national dengue fever/infectious diseases coordination mechanism (source: adapted from CBA5 presentation by T. Nguyen: www.slideshare.net/cbaiied/session-20-nguyen-vietnam-cca-health).

Note
Multi-level steering committees link Government of Vietnam and Vietnam Red Cross.

They can transmit community-level surveillance and risk information to higher-level authorities that influence resources and policy. This project in Vietnam demonstrates that government and NGO partnerships when joined up across national, subnational and community levels can meet community needs for the management and treatment of climate-sensitive diseases such as dengue fever.

Bangladesh: using radio and school curricula to communicate risks and empower communities with knowledge of risk preparedness

The location of Bangladesh between the Bay of Bengal and the Himalaya mountains makes this Deltaic region with three major river systems one of the most vulnerable countries in the world to the impacts of climate change. Bangladesh is frequently subject to the impacts of sea-level rise, salt-water intrusion of coastal regions, increased cyclones and flooding, riverbank erosion and land loss, as well as drought and erratic rainfall (MOEF 2008). Each of these environmental hazards poses a range of risks to the health, food security, water security and well-being of 140 million Bangladeshis, as well as increasing risks of displacement and migration (BCAS 2011). The first pillar of the National Climate Change Strategic Action Plan (MOEF 2008) is to 'protect the food security, social protection, and health' (MOEF 2008). To this end, the Ministry of Health and Family Welfare established a Climate Change and Health Promotion Unit (CCHPU)[3] to coordinate and implement risk reduction and adaptation research, monitoring and pilot adaptation measures.

Bangladesh has long concluded that community-based health interventions are cost-effective, achieve higher acceptability, allow for direct communication and rapid feedback with the most vulnerable, enable a reality check for both service recipients and service providers on needs and community choices and empower communities to act upon early warnings.[4] The Climate Change and Health Promotion Unit have been piloting several health adaptation projects to translate national risk management priorities to the community level. Mechanisms they have been using include: (1) community-based research on the sensitivity of health to climatic conditions (Khan *et al.* 2011); (2) community radio and 'edu-tainment' to communicate risks and transmit education about preparedness; (3) empowerment of youth as local activists, risk communicators and health advocates; (4) mainstreaming climate change risk awareness into school curricula to sensitise children to hazards and protective health behaviour; and (5) CBA using public–private partnership at vulnerable locations.

Two ways to communicate risks to communities and encourage protective behaviour and safety were attempted. First, they developed a school manual, *Climate Change and Health Safety*, to create awareness of climate risks to health among parents, students and teachers in Bangladesh. The manual was based on two WHO manuals for students and teachers (WHO-SEARO 2008). A range of stakeholders specialised in disease control and education, from government and civil society,[5] came together with experts in climate change and education manual development in order to incorporate the manual into the National School Curriculum.

Second, community radio has been effectively used in Bangladesh to educate the public on personal safety and preparedness, how to understand and respond to the severe weather early warning system and where and how to seek help during emergencies. The CCHPU of the Ministry of Health and Family Welfare, and two local NGOs of two different coastal areas at Kaliganj, Shatkhira (Radio Nolta) and Shitakundu, Chittagong (Radio Sagorgiri) have jointly developed two-hour-long radio programmes broadcast daily. These radio shows are an 'edu-tainment' approach, using folk media and local languages to reach vulnerable communities. Although the impacts of these programmes have not yet been evaluated, they aim to empower communities by increasing their awareness of health risks associated with climate change, the preventive actions they could take to stay healthy and how to understand and respond to various alerts issued by early warning systems for imminent threats of local floods and cyclones.

This case study is another example of how expanding partnerships to work with other sectors, such as education and the media, can enhance the reach of messages and information to communities, and scale-up CBA to protect health.

Jordan: integrated management of health risks from wastewater

Jordan is one of the most water-scarce countries in the world; water scarcity in Jordan is projected to worsen as a result of climate change (Ray *et al.* 2010). To adapt to future conditions and ensure adequate and safe water and food production, the nation of Jordan concluded that their national water sector policy requires modification (Alfarra *et al.* 2011). New national water management policies now allow for the increased use of wastewater, or reclaimed water from municipal sewage treatment facilities, to be used for irrigation in food production. Using reclaimed wastewater in Jordan not only carries cultural stigmas but also introduces new risks of diarrheal disease transmission. If reclaimed wastewater is not managed properly, diarrheal disease risks could increase for agricultural workers, their families and communities and the general consuming public of local agricultural products. Thus, to minimise these potential new health risks in Jordan, health authorities are working with the water and agriculture authorities to jointly develop a safe wastewater management system. They are beginning to employ the following actions:

- national regulations for microbial standards for food and water quality led by the Ministry of Health;
- national regulation of wastewater treatment and management procedures led by the water sector;
- managing risks associated with the reuse of treated wastewater requires a robust and integrated monitoring system. Three already existing systems need to be strengthened and synchronised to act as one supporting the reuse process: (1) close monitoring of wastewater effluents to make sure that both microbial and chemical characteristics of wastewater are within the new standards; (2) monitoring of crops that are produced by treated wastewater;

and (3) monitoring of health of the farmers, their families and the community at large;
- inclusion of farmers unions and farmers in discussion about policy changes and new 'field to market' crop monitoring that will to be carried out by the health, water, food and agricultural agencies;
- a commercial marketing campaign for the general public implementing 'safe food labels' which validate that foods have been grown with safe agricultural practices and monitored for quality, in order to gain social acceptance of new practices for using wastewater in agriculture; and
- a media campaign to manage risk perceptions and risk communication.

Implementation of wastewater safety protocols and a food safety management plan requires strong coordination and joint work plans among relevant agencies involved.

Figure 10.2 shows that national, district and community-level actors are needed for the implementation of national policy down to the community level. Key tasks such as food and water quality monitoring are conducted at community and district levels and involve farmers. A comprehensive mapping of districts and communities exposed to wastewater sources that may be at increased risk was conducted jointly between three government agencies to target vulnerable areas that should be prioritised for water and food safety monitoring.

Figure 10.2 Jordan: stakeholders in translation of national wastewater policy to the local level (source: adapted from CBA5 presentation by O.Kittaneh: www.slideshare.net/cbaiied/session-20-kittenah-jordan-cca-health).

Jordan identified that cultural acceptance and active buy-in from communities and farmers would be a critical aspect for this national water sector adaptation measure to be effective at the local level. This project developed advertising and marketing campaigns that engaged the national media to distribute messages to increase the social acceptance of the new water and agriculture policy and practice, and raise community awareness of wastewater safety measures. In addition, a consumer marketing instrument of 'health – labelling' of food products cultivated with reclaimed water will be trialled to build consumer confidence and promote food safety.

Discussion: 'scaling up' CBA to protect health

The three case studies describe a cooperative NGO and government approach to dengue control in Vietnam, a multi-sectoral and market-based approach to managing health risks of water stress in Jordan, and Bangladesh's approach to using the media and school curricula to sensitise and empower communities and children with information on what to do in the case of an emergency. These case studies illustrate innovative ways to scale-up traditional programming by joining up community-based interests and initiatives with top-down national policies, as well as collaborating across sectors, for the greatest protection of people's health.

Experience in CBA for health often highlights that the effectiveness, coverage and impact of interventions depends upon how effectively actors and actions that reach and empower communities are 'joined up'. Five common factors outlined in Table 10.1 emerge as important for successfully 'joining up' and 'scaling up' CBA in ways that can have the greatest reach and impact to protect community health from climate change.

The first factor useful for scaling up CBA is to communicate effectively and translate top-down policies and interventions set by national public health institutions to the community level, while also informing these interventions with bottom-up community-level action and realities (Ebi and Semenza 2008). Efforts to do so can improve the targeting and management of health risks associated with climate at the community level. Mechanisms such as tiered coordination structures and consultative processes can help in this translation process by creating opportunities for national policy-makers to listen to communities, acknowledge community perceptions of risk and priorities and communicate climate risks to communities in ways that empower local actors to reduce their own vulnerability in culturally appropriate ways. However, limited experience exists in how to present the risks of climate change (including uncertainties) at the community level and to convey that impacts may not occur if appropriate preventive action is taken (Red Cross Climate Centre, 2011). Poor communication of climate change risk can evoke a sense of despair, or an overwhelming perception of risks that may occur far in the future. Communicating uncertain risks to communities in a way that empowers communities to take action and reduce vulnerability is critical, because predictions of likely impacts at the community level are highly uncertain.

Table 10.1 Factors for scaling up CBA for health

	Success factor for scaling up CBA	*Purpose*	*Jordan*	*Bangladesh*	*Vietnam*
1	Communicate and translate top-down policies and interventions set by national public health institutions to the community level	To better target and manage health risks associated with climate	Yes	Yes	Yes
2	Collect community information about local health risks and complement with national and local epidemiological surveillance, as well as assessments and projections of global and regional climate and environmental conditions over time	To identify, prioritise and address community health risks and vulnerability in relation to the larger scale and longer-term climate risks in a given country	Yes	Yes	Yes
3	Incorporate health goals into the community-based development and adaptation actions of health-influencing sectors (e.g. water, agriculture, disaster management, education, energy)	To better protect communities from upstream health risks in other sectors	Yes	Yes	N/A
4	Intensify primary health care and preparedness efforts, particularly through the engagement and empowerment of actors, institutions and mechanisms that reach the community level	To reduce vulnerability and increase resilience to future climate change	Yes	Yes	Yes
5	Identify and utilise the institutions and mechanisms that link levels, sectors and actors who protect health and build resilience of communities	To facilitate, translate and actively communicate across levels, in order to keep community needs at the centre of development decisions	Yes	Yes	Yes

The second factor that can help scale-up CBA is to collect and complement community information about local health risks with national and local epidemiological surveillance, risk assessments, and projections of global and regional climate and environmental conditions over time. Efforts to do this can help identify, prioritise and address community health risks and vulnerability in relation to the larger-scale and longer-term climate risks in a given country.

Linking local observations of changes in health risk factors is essential to understanding the nature of potential future health risks for individual regions and communities. This is because some health risks may not be evident with a few observations in a community until observations are aggregated at higher levels (such as watershed or provincial level) to identify epidemiological trends or changing patterns in important risk factors over time and space. Additionally, in the case of climate change, local risk perceptions may only partially capture the magnitude or knowledge available on a particular risk and go beyond traditional knowledge and existing coping capacities. Aggregated perspectives of climate and environmental risks for community health are needed, with special attention given to the communication of risk magnitude, timing and uncertainty to community leaders.

Community-level information on vulnerability, capacity and local risks is also critical for strategic higher-level risk management and provision of support (Aalst *et al.* 2008). The use of organisations that reach the community level (such as NGOs/CBOs), innovative communication technology (e.g. mobile phone surveillance and public service messages) and coordination mechanisms can help collect and transmit information about communities to strategic decision-makers.

The third factor that can help scale-up CBA is to incorporate health goals into the community-based development and adaptation actions of sectors influencing health, such as water, agriculture, disaster management and energy. This will help protect communities from upstream health risks in other sectors. A 'health in all sectors' or cross-sectoral approach to health protection is of paramount importance to address the risks of climate change and a significant way to scale-up community health protection. Fundamental health determinants, such as adequate and safe water, food and shelter in many parts of the world, are expected to be significantly impacted by climate change and will have subsequent ramifications for community health (Confalonieri *et al.* 2007). In conjunction, the policy and resource management decisions being concurrently developed in water, agriculture, energy, transport, education, land management, and disaster management sectors can all have significant positive or negative impacts on the health and social well-being of communities. Scaling up health protection will also require 'scaling out' by mainstreaming health protection into the considerations and decision-making in other sectors that affect population health.

Information about changing conditions and status of core health determinants outside the health sector is also fundamental to decision-making and preparedness in the health sector. Health decision-makers need up-to-date information on changes in health risks such as water availability or quality, or safety of housing or infrastructure in relation to extreme events. Similar functions, mechanisms and institutions are involved in multi-sectoral partnerships. Common mechanisms include multi-sectoral assessments and decision-teams, policy advocacy and joined up monitoring (Rehfuess *et al.* 2009). Multi-sectoral early warning systems and coordination are also critical mechanisms to link sectors and to communicate and respond to community health risks and needs.

Addressing health risks at the community level often requires an integrated approach. Sectoral divisions are not so obvious at a community or household level. This is because community members are not likely to create distinctions in sector-specific interventions the way governments, NGOs or academics do. Additionally, when health protection is the goal and guiding principle in a community, combined planning and execution of policy or action in water and sanitation, livelihoods, disaster risk reduction, as well as primary health care, commonly naturally emerge. These types of integrated risk management opportunities often become clear during community-based assessments, mapping or planning. As the multifaceted challenges of climate change makes cross-sectoral integration even more pertinent, top-down programming and responses should take advantage of the natural integration that occurs at the community level.

The fourth factor that can help scale-up CBA is to intensify community health and primary health care efforts along with community preparedness mechanisms, particularly through the engagement and empowerment of community actors, institutions and mechanisms. Efforts to manage and reduce current vulnerability, such as maintaining vaccination coverage, actively managing infectious diseases, and empowering networks of community health workers as well as youth, will significantly help communities respond to the impacts of future climate change.

Community health workers are key actors for intensifying community health, as they are responsible for the delivery of many key health functions at the community level and often drive critical outreach activities for personal and community safety, communication and response to emergencies. Youth can also be important actors for risk communication and community advocacy. Youth can play an important role in the community to identify and reduce health risks, such as being empowered to identify breeding sites of disease-carrying mosquitoes. Empowering youth and children with knowledge about risks, what to do in the case of an emergency and personal safety and prevention of risk can help protect the whole family and community.

The fifth factor that can help scale-up CBA is to empower and invest in the institutions and mechanisms that serve to join-up top-down and bottom-up efforts, transmit information across levels, connect multi-sectoral actors and generally extend the reach of programmes and communication to and from communities themselves. The scope of activities these institutions and mechanisms carry out often range across areas of research and risk identification, translation of science, identification of relevant cultural beliefs and perceptions, risk monitoring, risk communication, preparedness, health promotion and health services delivery, training and skill building, and risk management and response. The wide variety of mechanisms and institutions working at the community level serve common and critical functions as 'interlocutors' to enhance coordination, reach, efficacy and local engagement.

For example, mechanisms that facilitate dialogue and link community perspectives with national climate and risk information are important entities to shift actions from conventional development, to be appropriately climate informed and 'climate proofed'. It is in such coordination mechanisms, such as in the case study from Vietnam, where the place and space for 'joining up' can occur and enhance the effectiveness of multi-scale partnerships by sharing surveillance data with public health actors. NGOs, academia and other institutions are key to transmitting and translating this information, but need financial resources to complement government processes. Strengthening these mechanisms that translate and actively communicate across levels and actors can enhance strategic planning, resource mobilisation and scaling up health protection are essential to keep community needs at the centre of development decisions.

Another important opportunity for CBA is to build upon the well-established procedures and mechanisms for local health monitoring and disease surveillance that often exist at the community level. These local surveillance systems can be enhanced to be more sensitive and attentive to changing trends and patterns in health risks associated with climate change and a vital opportunity to monitor community vulnerability. NGOs and CBOs, and other agencies that work at the community level, serve as critical institutions for collecting and joining this local information with national-level information and policy. Because of their flexibility and unique reach to vulnerable populations, NGOs and CBOs can often provide valuable information about community-level vulnerabilities for assessments and fill gaps in government service coverage for delivery of information, education or primary health services.

It is widely recognised that improving surveillance systems is key to the detection of changing patterns of disease due to climate variability and climate change; however, multiple actors across scales are needed for surveillance to effectively work. Epidemiological surveillance, as well as adaptation planning and coordination, are actions generally coordinated by central government health authorities in most countries. NGOs and CBOs working in partnership with governments can effectively increase coverage of these functions in a country. Moreover, NGOs and CBOs rely upon national government surveillance data for information such as the distribution of vectors or cases of cholera for effective targeting of interventions, and are commonly restricted by the quality and availability of this data to monitor population vulnerability and changing disease risks. In partnership with government, NGOs and CBOs can be instrumental suppliers and quality controllers of surveillance information into the formal system from remote areas or other hard-to-reach populations who are vulnerable to climate change impacts.

Conclusions

Community health in developing countries is highly vulnerable to climate change. Adapting to climate change to maintain healthy communities will require strong community engagement and highly strategic interventions and mechanisms to bring together the strengths and competencies of top-down and bottom-up community efforts. Communities in developing countries are a resource to be catalysed, but need support to identify and understand risks, make decisions, and access adequate resources and options to be able to respond to climate risks. Community health and primary health care practices provide a strong foundation to do this, and lend significant strengths to efforts for adaptation to climate change.

This chapter presented five factors important for the scale-up of CBA for health protection.

1 To effectively implement national policies and multi-sectoral adaptation initiatives at the local level, top-down policy-makers must listen to communities, empower local actors to reduce their own vulnerability in culturally appropriate ways, communicate risk effectively and acknowledge community priorities and risk perceptions.
2 To prioritise and address community health with respect to the larger-scale and longer-term climate risks in a country, then community information about local health risks must be complemented and informed by national and local epidemiological surveillance, and assessments and projections of how regional climate and environmental conditions may change over time.
3 To prevent and protect communities from upstream health risks, health goals should be incorporated into all sectors, particularly water, agriculture, disaster management and energy.
4 To reduce vulnerability to future climate change it is essential to intensify community health and primary health care services.
5 To keep community interests and needs at the centre of development and adaptation policy and practice, intermediary institutions and mechanisms that link community to national levels and work across sectors and actors should be engaged and empowered.

Investment in these five areas will be critical for scale-up of CBA that protects human health.

Notes

1 www.asiaone.com/Health/News/Story/A1Story20101012–241839.html.
2 This project, 'Health Risk Management in a Changing Climate', is supported by a grant from the Rockefeller Foundation.
3 Bangladesh Climate Change and Health Promotion Unit website: www.cchpu-mohfw.gov.bd.
4 Opinion of the Bangladesh Ministry of Health and Family Planning: www.slideshare.net/cbaiied/session-20-kabir-bangladesh-cc-human-health.

5 Representatives from the Ministry of Education, Ministry of Environment and Forest, Ministry of Health and Family Welfare (MOHFW), DGHS, National Institute of Preventive and Social Medicine (NIPSOM), IPH, Institute of Epidemiology Disease Control and Research (IEDCR), International Centre for Diarrheal Disease Control Bangladesh (ICDDRB), BHE, NGO, civil society and Member and Senior Specialists of the National Curriculum and Text Board (NCTB) of the MOE.

References

Aalst, Maarten K. van, Terry Cannon and Ian Burton. 2008. 'Community Level Adaptation to Climate Change: The Potential Role of Participatory Community Risk Assessment'. *Global Environmental Change* 18 (1): 165–179.

Alfarra, Amani, Eric Kemp-Benedict, Heinz Hötzl, Nayif Sader and Ben Sonneveld. 2011. 'A Framework for Wastewater Reuse in Jordan: Utilizing a Modified Wastewater Reuse Index'. *Water Resources Management* 25 (4): 1153–1167.

Benitez, M.A. 2009. 'Climate Change Could Affect Mosquito-borne Diseases in Asia'. *Lancet* 373 (9669): 1070.

Confalonieri, U., B. Menne, R. Akhtar, K.L. Ebi, M. Hauengue, R.S. Kovats, B. Revich and A. Woodward. 2007. 'Human Health'. In *Climate Change 2007: Impacts, Adaptation and Vulnerability. Contribution of Working Group II to the Fourth Assessment Report of the Intergovernmental Panel on Climate Change*, 391–431. Cambridge: Cambridge University Press.

Costello, A., M. Abbas, A. Allen, S. Ball, S. Bell, R. Bellamy, S. Friel, *et al.* 2009. 'Managing the Health Effects of Climate Change: Lancet and University College London Institute for Global Health Commission'. *Lancet* 373 (9676): 1693–1733.

Dodman, David, and Diana Mitlin. 2011. 'Challenges for Community Based Adaptation: Discovering the Potential for Transformation.' *Journal of International Development.* doi:10.1002/jid.1772. http://onlinelibrary.wiley.com/doi/10.1002/jid.1772/abstract.

Ebi, K.L. and Jan C. Semenza. 2008. 'Community-based Adaptation to the Health Impacts of Climate Change'. *American Journal of Preventive Medicine* 35 (5): 501–507.

ExpandNet. 2011. 'ExpandNet.' www.expandnet.net/tools.htm (accessed 4 October 2011)

Haines, Andy, David Sanders, Uta Lehmann, Alexander K. Rowe, Joy E. Lawn, Steve Jan, Damian G. Walker and Zulfiqar Bhutta. 2007. 'Achieving Child Survival Goals: Potential Contribution of Community Health Workers.' *Lancet* 369 (9579): 2121–2131.

Hanson, Kara, M. Kent Ranson, Valeria Oliveira-Cruz and Anne Mills. 2003. 'Expanding Access to Priority Health Interventions: A Framework for Understanding the Constraints to Scaling-up'. *Journal of International Development* 15 (1): 1–14.

Keim, Mark E. 2008. 'Building Human Resilience: The Role of Public Health Preparedness and Response as an Adaptation to Climate Change'. *American Journal of Preventive Medicine* 35 (5): 508–516.

Khan, Aneire Ehmar, Andrew Ireson, Sari Kovats, Sontosh Kumar Mojumder, Amirul Khusru, Atiq Rahman and Paolo Vineis. 2011. 'Drinking Water Salinity and Maternal Health in Coastal Bangladesh: Implications of Climate Change'. *Environmental Health Perspectives* 119: 1328–1332.

MOEF. 2008. *Bangladesh Climate Change Strategy and Action Plan. Government of the People's Republic of Bangladesh: Ministry of Environment and Forests.* www.google.com/firefox?client=firefox-a&rls=org.mozilla:en-US:official (accessed 6 July 2011) .

Neira, M., R. Bertollini, D. Campbell-Lendrum and D.L. Heymann. 2008. 'The Year 2008: A Breakthrough Year for Health Protection from Climate Change?' *American Journal of Preventive Medicine* 35 (5): 424–425.

Nerlander, Lina, Tu Ngoc Chau and Thuan Nguyen. 2011. *Study on Knowledge, Attitudes and Practices of Community Members in Tien Giang and Ho Chi Minh City with Regards to Dengue Fever and Climate Change*. Hanoi: Red Cross Climate Centre.

Ray, P.A., P.H. Kirshen, D.E. Rosenberg and R.E. Hagan. 2010. 'An Overview of the Water Situation in Amman, Jordan.' *ASCE Conference Proceedings* 371 (41114): 207–207.

Red Cross Climate Centre. 2010. 'Health Risk Management in a Changing Climate: A Global Approach to Building Local Capacity.' www.climatecentre.org/downloads/File/programs/Rockefeller%20short%20project%20description%20for%20website.pdf (accessed 29 June 2011).

Red Cross Red Crescent Climate Centre. 2011. 'Health Risk Management in a Changing Climate: An Operational Research on Dengue Fever Prevention in Vietnam.' www.climatecentre.org/downloads/File/asia_and_pacific/CBA5%20RC%20flyer.pdf.

Rehfuess, Eva A., Nigel Bruce and Jamie K. Bartram. 2009. 'More Health for Your Buck: Health Sector Functions to Secure Environmental Health.' *Bulletin of the World Health Organization* 87 (11): 880–882.

Syed, Shamsuzzoha B., Adnan A. Hyder, Gerald Bloom, Sandhya Sundaram, Abbas Bhuiya, Zhang Zhenzhong, Barun Kanjilal, *et al.* 2008. 'Exploring Evidence–Policy Linkages in Health Research Plans: A Case Study from Six Countries'. *Health Research Policy and Systems* 6: 4.

WHO. 2008. *Climate Change and Health: Report by the Secretariat to the 124th Session of the Executive Board of the World Health Organization.* http://apps.who.int/gb/ebwha/pdf_files/EB124/B124_11-en.pdf (accessed 8 June 2011).

WHO. 2009. *Practical Guidance for Scaling up Health Service Innovations*. www.who.int/reproductivehealth/publications/strategic_approach/9789241598521/en/index.html.

WHO-SEARO. 2008. *How Is Climate Change Affecting Our Health: A Manual for Teachers*. www.who.int/globalchange/training/en (accessed 8 June 2011).

11 'Down home, it's all the same'

Building synergisms between community-based disaster risk reduction and community-based climate change adaptation

Ben Wisner, Marcus Oxley, Pramita Harjati Budihardjo, Karina Copen, Gina Castillo, Terry Cannon, Jessica Mercer and Stephane Bonduelle

The challenge and our contribution

Except for places that are in the throes of civil war, most ordinary rural and urban dwellers live lives punctuated by the risks and opportunities that occasionally interrupt the rhythms of work and rest. Of course, widespread violence does not preclude the added burden of natural hazards and makes coping and response more difficult (Wisner 2012). Also in the United States and elsewhere, although governments may emphasise the primacy of terrorism, quotidian risks prevail: hurricane, tornado, earthquake, flood, drought and wildfire (Knowles 2011; Perrow 2007).

Over the past two decades some very effective approaches have been developed that invite and support groups of people living in the same localities to assess the specific natural hazards they face, the people and assets at greatest risk and also to appraise their own resources and capacities to reduce losses or even to prevent the hazard. This suite of methods has become known as community-based disaster risk reduction (CBDRR).

Over this same two decades, climate science has made it clearer and clearer that global climate change has specific, localised impacts on the frequency and severity of climate-related natural hazards (IPCC 2012). As global discussion of reducing climate-changing emissions ('mitigation') was complemented with the notion of living with and adjusting to climate change ('adaptation'), some of the same international non-governmental organisations (INGOs) that had been successful with CBDRR began to experiment with a similar, climate-focused approach: community-based climate adaptation (CBA).

The challenge, therefore, is to clarify how best CBDRR and CBA can be integrated into a cost-effective and effective extension and local mobilisation approach, and how the results can be up-scaled as inputs to territorial and national planning. This chapter attempts such a clarification.

Historical context

In May 1994, at the mid-point of the United Nations Decade for Disaster Risk Reduction (IDNDR), several international organisations such as the International Red Cross/Crescent Federation (IFRC) and civil society organisations asked for a 'time out'. They noted that community participation had played a very small part in the first half of the IDNDR. The mid-term consultation took note, and one could (arbitrarily) date the mainstreaming of communities as partners in disaster reduction from that meeting. Years later, there now exists a Global Network of Civil Society Organisations for Disaster Reduction (GNDR), and most governments and international NGOs and UN organisations explicitly recognise CBDRR as one of the pillars of human security (see Delica-Willison and Gaillard 2012; Thompson 2012). This shift came after another sea change in the approach to natural hazards and disaster risk. From the early 1980s, the earlier physicalist/technocratic focus on the triggering hazards had been effectively challenged, expanded and balanced with an equal concern for the vulnerability and capacity of people (Cuny 1983; Hewitt 1983; Anderson and Woodrow 1988; Maskrey 1987).

A strikingly similar evolution of ideas and practice has also attended the deepening of 'climate change adaptation' (CCA) by focusing on what communities actually do – with or without outside encouragement and support to reduce the potential impacts of climate change on their livelihoods ('autonomous adaptation' and 'community-based adaptation'). However, a major difference is that CBA at the time of writing remains an accepted part only of INGO and civil society practice, with some minor uptake by some UN agencies and government agencies (e.g. in Bangladesh). For the most part, national governments in mid- and low-income countries are so bound up by their top-down National Adaptation Programmes of Action (NAPAs) and live in such high hopes of large flows of top-down funding for 'adaptation', that they have maintained tight control of discourse of climate change and have given little attention to what people are actually doing. For example, in Tanzania, government CCA policy is tightly coupled with ambitious plans to introduce genetically modified crops (maize, in particular) and to create and mechanise large farms with the assistance of overseas direct investment by agribusiness corporations. The approach would blanket a country the size of France and Germany combined, where people speak 122 vernacular languages, eat many different staple foods and dwell in 12 different climatic regimes with a single, one-size-fits-all plan (Wisner *et al.* 2012).

In this light, there are two sets of benefits from clarifying and, above all, programming on the basis of, synergisms between CBDRR and CBA. The first set of benefits has to do with moving the focus of national governments in the direction of communities, local knowledge and capacity. This, however, is part of a broader agenda to do with improving governance, fighting corruption, and fully implementing decentralisation.

The second set of benefits concern ongoing efforts by some governments and many INGOs and civil society organisations (CSOs) that already understand and accept the importance of community-based capacity and action. These include encouragement of collaboration and system thinking, optimisation of human and financial resources, and allowing fine-tuned, highly localised action in response to details of natural and social conditions. As Shaw *et al.* (2010: xix) put it, 'synergies between DRR and CCA would be necessary not only to avoid duplication and derive optimal benefits from scarce resources, but also to add value to the projects from the lessons learnt from the respective perspectives'.

Opportunities and basis for synergism

We finish these introductory remarks by highlighting a number of facts that suggest a prima facie case for integrating CBDRR and CBA in order to attain the synergistic benefits mentioned earlier. Note, for example, that:

1 Ninety per cent of all natural hazards DRR and CBDRR deal with are climate related (all except earthquake, tsunami and volcanic eruption).
2 Local urban and rural dwellers experience poverty, disasters and climate change, and manage risks and opportunities, holistically; as the chapter title says, 'down home it's all the same'.
3 The same local government and civil society personnel end up required to deal with both CCA and DRR.
4 Both CBA and CBDRR advocate for similar approaches that are people-centred, strengthen local capacities and adopt participatory methodologies (i.e. participatory assessment, planning, implementation, monitoring and evaluation).
5 Both CBA and CBDRR are focused on the protection of lives, livelihoods and assets, and both can easily be seen as aspects of sustainable human development (risk-aware development/climate-smart development).
6 Common frameworks can increase potential for collaborative actions and coherent programming and are less confusing and demanding for vulnerable people and local institutions in practice.

Review of available frameworks that link CBDRR and CBA

It is convenient to distinguish analytical from policy frameworks for linking CBDRR and CBA, although they may overlap. Analytical frameworks seek to guide empirical research in order to understand the mechanisms and functional connections that link by cause and effect behaviours that fall into the category of CBDRR, outcomes for CBA and vice versa. For example, terracing hillside farms in the wet-dry tropics conserves water in the drier season, prevents landslides and increases production. The empirical connections work in several directions. Typically, such analytical frameworks focus on the micro-scale: household livelihood systems, community decisions and actions and activities at the scale of local government and local civil society. The reason for this is that whether or not national government policy exists or is implemented at the local scale, people perceive risks and opportunities, including those accompanying climate change, and they modify their behaviour. This is referred to as 'autonomous adaptation'.

Analytical frameworks

Analytical frameworks embed the flagging up of micro-scale processes in visual representations of macro process, some even at the global scale. A common problem with such frameworks is that their application requires researchers to use multiple methods, some of which are quantitative and some qualitative: historical, political economic, socio-economic, ethnographic and even linguistic. Under the dominance the Intergovernmental Panel on Climate Change (IPCC), 'climate science' privileges quantitative evidence and the use of a hypothetico-deductive paradigm. Positivism rules climate science. Given the origin of climate science in climatology, atmospheric physics and mathematical modelling, such a definition of 'evidence' is understandable. However, this makes it difficult for the sort of evidence collected, for example, by INGOs, NGOs and CSOs using participatory action research, to gain acceptance by mainstream climate science. We return to this problem below.

Perhaps the most ambitious analytical framework is that offered in the *South Asia Disaster Report 2010* produced by Practical Action and Duryog Nivaran (2011) and adapted for the African context by Wangui *et al.* (2012) (Figure 11.1).

While the central focus of this framework is the lack of 'adaptive capacity' – a core concept for researchers and practitioners dealing with CCA and CBA and also 'hazard', 'disaster' and 'recovery', which are core concepts for workers in the field of DRR, including CBDRR – the context is far broader. This larger context is global and national, and in both cases the analytical framework designates perverse incentives and misuse of power as areas requiring close research attention. One should recall that analytical frameworks are not models of how the world has been shown definitively to work, but guides to how the world seems to work – thus a sort of vade mecum for the researcher who has limited time and resources and needs to focus on the most important questions (Wisner *et al.* 2012b). As such, Figure 11.1 serves the same 'framing' function as the better known Pressure and Release Model (Wisner *et al.* 2004).

'Maldevelopment' (*maldéveloppement*) is a term coined by Samir Amin, a critical historian and development economist, an Egyptian resident for many years in Senegal (Amin 2011). The term's invention was a way of calling attention to the way that the terms 'developing country' and 'underdevelopment' had been for too long accepted uncritically. If the world does, indeed, have a single, complexly interconnected economy – a fact now even more evident than in Amin's day, then might not 'underdevelopment' in the global South be only one aspect of the dysfunction of the global system? Likewise, 'induced' development has been observed for a long time. Colonial powers were sometimes quite brutal as they 'induced' local people to produce for the market in Belgian, Portuguese and Dutch colonial territories. Today, induced development takes the form of

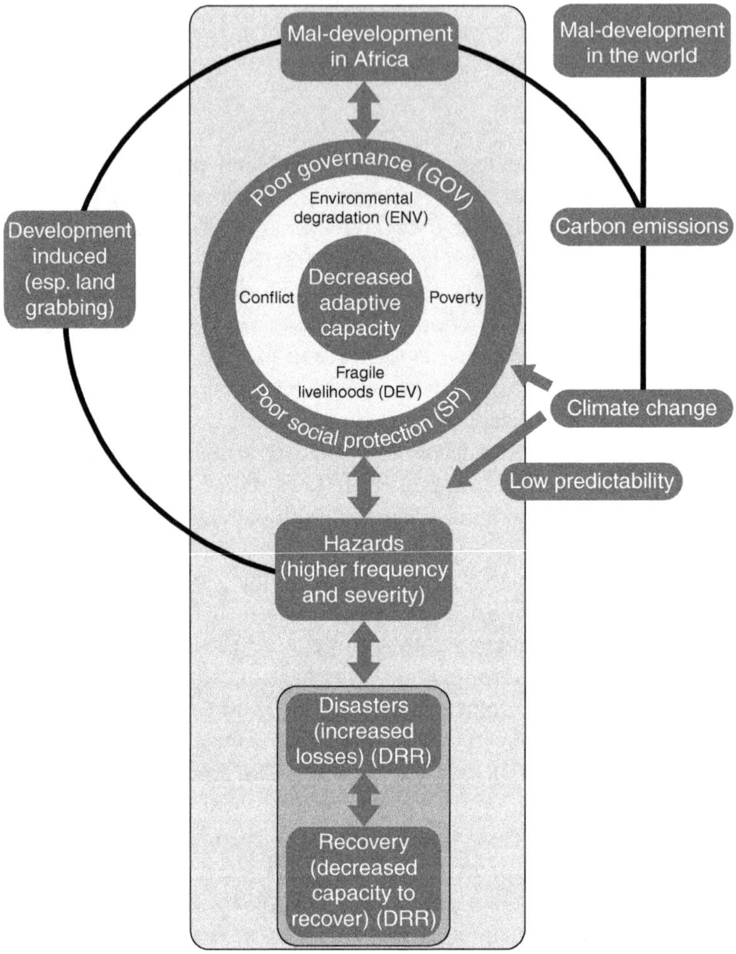

Figure 11.1 Analytical framework from *South Asia Disaster Report 2010*, adapted for Africa.

direct overseas investment with the blessing of national governments. In the course of these investments, land is often appropriated for large-scale production of bio-fuels (a link to the climate and carbon discourse), food for export or as hunting or forest reserves and plantations (World Bank 2011).

The value of Figure 11.1 as an analytical framework is to focus the attention of researchers on the root causes and dynamic pressures that underlie the vulnerability of various groups of people to both disaster risk and to the threats of climate change (Wisner *et al.* 2004).

Table 11.1 (Oxley 2011) is a more detailed analytical framework that takes into account five sets of key determinants that affect resilience against both

Table 11.1 Marcus Oxley's/GNDR's draft core determinants of resilience shared by DRR and CBA

1 Governance (sub-components)	*2 Knowledge (sub-components)*	*3 Sustainable livelihoods (sub-components)*	*4 Built environment (sub-components)*	*5 Natural environment (sub-components)*
• Decentralised authority and devolved responsibilities • Participatory decision making and inclusive planning • Contingencies/emergency preparedness planning • Representation/inclusion of the poor • Gender perspectives • Incentivise multi-stakeholder partnerships • Relevant policies and legislation • Institutional/functional capacities and structures • Access to financial resources • Human resources/skills base • Public accountability • Participatory policy monitoring • Enforcement mechanisms • Transparency – access to relevant risk information • Coordination capabilities • Collaborative approaches • Trust building (social capital) • Strategic leadership	• Periodic participatory risk assessment and mapping • Climate models and forecasts • Early warning systems • Linking local indigenous and scientific knowledge • Access to climate information • Formal and informal education and training • Knowledge innovation, creation and management • Community of practitioners • Community–peer learning • Examples of good practice • Unifying frameworks • Public awareness and perceptions of acceptable risk • Public awareness of rights and responsibilities • Public 'demand' for action • Multi/social media usage • Experimental, iterative, learning approaches • Capacity to gather, analyse data, formulate responses	• Capacities to secure basic needs and living conditions • Access to information, productive resources and assets • Access to common pool resources • Adequate and fair wages • Access to markets • Diversified livelihood/potential future options • Early warning/early action contingencies, preparedness • Secure land tenure and property rights • Savings and credit • Micro-insurance schemes • Reserves, substitutable assets • Social protection schemes • Poverty alleviation schemes • Health and social services • Equitable economic development and distribution of livelihood assets	• Land use planning and building codes • Integrity and location of community facilities (fire, police, admin, health) • Design, location, open space • Housing design and construction (formal/informal) • Neighbourhood design and location (formal/informal) • Emergency management facilities; e.g. evacuation/access routes, shelter • Local disaster plans (preparedness and response) • Resilient private and public transportation systems • Utilities (water, sewerage, electricity, communications)	• Geological, hydrological, ecological assessment and mapping (EIA) • Understanding of functioning, capacities of natural systems • Relationship between livelihood systems and natural systems • Relationship between built environment and natural systems • Policy, legislation and practices to support natural resource management • Application of indigenous knowledge and practices • Low carbon development (mitigation co-benefits) • Renewable energy sources • Protection of indigenous peoples' rights and entitlements • Restore, sustain and develop capacity of ecosystems

climate change and natural hazards: governance, knowledge, livelihood characteristics, the built environment and the natural environment. Nine to 18 sub-components are identified under each of the five core determinants.

Again, as with all analytical frameworks, the components and sub-components are suggested to researchers as useful foci for investigation. They are found in the framework because of the empirical results of both past research and less formal observation by practitioners of CBDRR and CBA.

The notion of 'resilience' originates in ecology, engineering and materials science. It had the meaning of 'bouncing back' after a stress or continued functionality. It was adopted by climate science to refer to the ability of human systems at many scales (household livelihoods, urban infrastructure, national economies) to 'bounce back' after the stress of climate change. In its use by scholars and practitioners of both CBA and CBDRR, the meaning has shifted to include transformational change, or 'bouncing forward'. Successful adaptation to climate change and successful DRR require change at various scales. Applied to Figure 11.1, the notion of transformational change would suggest that, among other things, 'maldevelopment' must change – a complex, difficult political-economic process. More modestly, land use, location, ways of making a living can also transform themselves as long as the society in question provides access to those at risk to the new locations and resources necessary for transformation.

Key determinants of resilience

Policy frameworks

Policy frameworks adopt or assume definitions and relationships developed through analysis. Policy frameworks assume the empirical validity of these relations, unlike analytical frameworks which do not assume absolute truth but encourage or guide further research. This remark is not meant as a critique of policy frameworks. In the end, policy must be made on the best available knowledge at a point in time. The best policy frameworks are, of course, open to iterative improvement as research and analysis progress. International and government action, the links among institutions and direction of funding are laid out by such policy frameworks. Frameworks are used by those who control finance and other resources, as well as those who seek to advocate and influence the use of resources. These frameworks are also normative once adopted by an institution. An ambitious example is provided in Figure 11.2 (Wiggins 2011: 3), where the goals of full horizontal and vertical integration of all work on adaptation (including, one must assume, CBA and CBDRR) are suggested.

While Figure 11.2 is quite general, Figure 11.3 (Mitchell *et al.* 2010: 4) lays out a series of detailed imperatives for integration under three headings. The first set of policy imperatives this checklist highlights is to 'tackle changing disaster risks and uncertainties'. Here the focus is on the physical processes that link climate change and disaster risk, as well as to monitor how changes affect

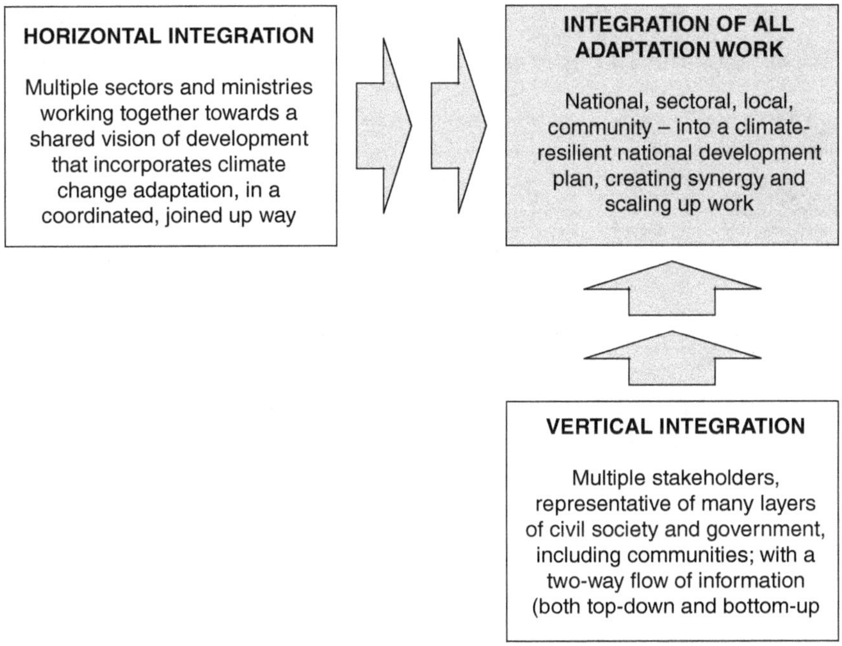

Figure 11.2 Adaption united framework.

people's livelihoods. The second set of imperatives concern enhancement of 'adaptive capacity'. The emphasis is on social learning, tools and policies that facilitate learning and the use of tools, especially at the local scale. The third column is striking and provides a link from the domain of this framework back to that of the first analytical framework presented above. Here, the policy framework encourages the reader to 'assess poverty and vulnerability and their structural causes'. Such broad and comprehensive framing of policy begins, if taken seriously, to get at the root causes underlying social vulnerability to climate-related hazards.

Strengths shared by CBDRR and CBA and some differences

Political will

CBDRR benefits from moderate national political will generated over nearly 25 years through the work of the IDNDR (1990–1999), the coordinating and catalysing activities of the UN International Strategy for Disaster Reduction (2000 onwards) and, alas, experience in most countries of deadly and costly disasters during this period. However, political will supporting climate change adaptation is backed up internationally by treaty, as the UNFCCC and work of the IPCC has increasingly focused on 'adaptation' as the counterpart of 'mitigation'. CBA,

1. Tackle changing disaster risks and uncertainties

1a

Strengthen collaboration and integration between diverse stakeholders working on disasters, climate and development

To what extent are climate change adaptation, disaster risk management and development integrated across sectors and scales? How are organisations working on disasters, climate change and development collaborating?

1b

Periodically assess the effects of climate change on current and future disaster risks and uncertainties

How is knowledge from meteorology, climatology, social science, and communities about hazards, vulnerabilities and uncertainties being collected, integrated and used at different scales?

1c

Integrate knowledge of changing risks and uncertainties into planning, policy and programme design to reduce the vulnerability and exposure of people's lives and livelihoods

How is knowledge about changing disaster risks being incorporated into and acted upon within interventions? How are measures to tackle uncertainty being considered in these processes? How are these processes strengthening partnerships between communities, governments and other stakeholders?

1d

Increase access of all stakeholders to information and support services concerning changing disaster risks, uncertainties and broader climate impacts

How are varied educational approaches, early warning systems, media and community-led public awareness programmes supporting increased access to information and related support services?

2. Enhance adaptive capacity

2a

Strengthen the ability of people, organisations and networks to experiment and innovate

How are the institutions, organisations and communities involved in tackling changing disaster risks and uncertainties creating and strengthening opportunities to innovate and experiment?

2b

Promote regular learning and reflection to improve the implementation of policies and practices

Have disaster risk management policies and practices been changed as a result of reflection and learning-by-doing? Is there a process in place for information and learning to flow from communities to organisations and vice versa?

2c

Ensure policies and practices to tackle changing disaster risk are flexible, integrated across sectors and scale and have regular feedback loops

What are the links between people and organisations working to reduce changing disaster risks and uncertainties at community, sub-national, national and international levels? How flexible, accountable and transparent are these people and organisations?

2d

Use tools and methods to plan for uncertainty and unexpected events

What processes are in place to support governments, communities and other stakeholders to effectively manage the uncertainties related to climate change? How are findings from scenario planning exercises and climate-sensitive vulnerability assessments being integrated into existing strategies?

3. Address poverty and vulnerability and their structural causes

3a

Promote more socially just and equitable economic systems

How are interventions challenging injustice and exclusion and providing equitable access to sustainable livelihood opportunities? Have climate change impacts been considered and integrated into these interventions?

3b

Forge partnerships to ensure the rights and entitlements of people to access basic services, productive assets and common property resources

What networks and alliances are in place to advocate for the rights and entitlements of people to access basic services, productive assets and common property resources?

3c

Empower communities and local authorities to influence the decisions of national governments, NGOs, international and private sector organisations and to promote accountability and transparency

To what extent are decision-making structures de-centralised, participatory and inclusive? How do communities, including women, children and other marginalised groups, influence decisions? How do they hold government and other organisations to account?

3d

Promote environmentally sensitive and climate smart development

How are environmental impact assessments including climate change? How are development interventions, including ecosystem-based approaches, protecting and restoring the environment and addressing poverty and vulnerability? To what extent are the mitigation of greenhouse gases and low emissions strategies being integrated within development plans?

Figure 11.3 CLIMATE SMART disaster risk management framework.

while a more recent interpretation of what effective and people-centred 'adaptation' must look like, nevertheless enjoys by association the political will generated by so much diplomatic work and investment in climate change. By contrast, CBDRR has not yet been able to generate the same level of national political will and is not backed up by a legally binding treaty (Pelling and Schipper 2009).

Analytical complementariness

CBA has a strong analysis of the role of local ecosystems in the emergence of future climate-related hazards and in local options for adapting to them. Linked to a livelihood analysis, CBA yields practical and often low-cost actions that can be taken even in low-income communities, as long as livelihoods are simultaneously enhanced. CBDRR is supported by a strong analysis of some of the proximate, ecosystem and livelihood-based risk factors that CBA excels in mapping, but until now it has been relatively weak in understanding the physical and biological nature of hazards. CBDRR's theoretical basis has often been biased towards root causes of social vulnerability/marginality such as poor governance, misuse of economic and political power and discrimination against minority groups. Thus the combination of CBA and CBDRR at the local scale provides communities, civil society and local government with guidance to actions required to enhance resilience (in the sense of transformational change) in the face of a wide range of hazards, including climate-related ones.

Timescale

CBA is concerned both with how people adapt to gradual changes in climate *means* (averages) and how they can protect themselves, settlements and livelihoods from climate-related *extremes*. Certainly, CBA may benefit from understanding what preoccupies CBDRR, namely how people 'cope' with climatic-related hazards (using the term more common in DRR). Less obvious, but true, is the benefit CBDRR can derive from putting its understanding and practice in the longer historical timescale of people's adaptation to climate change. While true that CBDRR must confront short onset hazards such as landslides and earthquakes, some, usually with moderate timescales such as volcanic eruptions and their seasonal aftermaths (lahars), and slow onset hazards such as drought, of these time scales are shorter than the anticipation of long-term sea-level rise, etc. (typically out to the year 2050 or even 2100).

Challenges and obstacles to synergy

Some of the challenges are well-known, recurrent ones. These include competition among INGOs, academic researchers and government ministries/departments for funding and lack of understanding of the benefits of synergy on the part of donors/grant-makers/government officials in charge of budget allocation. Another is stove piping of legislation, policy and programmes that arise and evolve in parallel whenever there are large sums of money available from donors for one or another major 'flavour of the decade', such as the Millennium Development Goals or climate change.

Other obstacles have to do with common features of national governments. These include public administration cultures that foster specialist silos and separate, parallel institutional structures for 'sectors' (e.g. ministries or departments of X,Y, Z). This, in turn, is sometimes the excuse for and certainly exacerbates a persistent and all too common gap between national and local in terms of implementation, especially as regards decentralisation of financial resources (GNDR 2011; UNISDR 2011). Birkman and Teichman (2010) refer to this problem as functional scale mismatch.

Despite the complementarities discussed earlier, there are also what Birkman and Teichman call spatial and temporal scale mismatches. The key sides to IPCC and UNFCCC work, 'mitigation' and 'adaptation', can actually overlap as sets of activities. For example, preservation of natural forest regulates water supply downstream and helps prevent landslides and flooding ('adaptation'), while it also locks up carbon ('mitigation') – a fact recognised by the funding system known as Reduction of Emissions from Deforestation and Forest Degradation (REDD). As such, some CCA initiatives, including community-based ones, may encompass a whole watershed. CBDRR has tended so far to be more localised in specific communities.

Temporal mismatch is another challenge. As noted earlier, even slow onset hazards such as drought involve timescales much shorter than those involving the major changes implied by warming of 3–4°C or even higher. On the other hand, to some degree CBDRR and CBA remain stuck in the near past, present and near future. Both would benefit from a much longer-term perspective. Much CBA does not address the longer-term future and is really only reactive to current perceived or actual climate change, not proactive.

For example, on the borders of the Sunderbans in coastal Bangladesh, cyclone recovery has involved innovative, participatory house and community building design as well as highly efficient use of polyculture and fishing around ponds and planting on dikes (UNEP 2009). However, none of this CBA takes into account projections that would see this entire zone underwater by 2050. One has to ask if such a timescale is too much to ask of farmers, shrimp farm workers and fishers – not because they are 'uneducated' or 'not experts', but because they have no resources to consider moving inland, and nowhere to go. Might a savings and investment fund be developed by one of the highly competent large micro-credit organisations such as BRAC or Gramin Bank? Making small monthly deposits, these working poor could invest in a fund that would allow their grandchildren a chance to move. However, should they? The downscaling of climate change models is difficult, and at the local scale there is great uncertainty as to specific physical and biological impacts and the timescales involved (Puhlin *et al.* 2010: 460). The poorest people in the world, those living on less than $2 per day, for example, are likely not to opt for long-term adaptations until (1) their living standard increases and (2) the local interpretation of climate models is more certain.

Case studies

Rural CBA/CBDRR in El Salvador (Oxfam 2011)

The 'Reducing Vulnerabilities in Ahuachapán and Sonsonate' (PRVAS) project was designed in 2005 after Hurricane Stan and the eruption of the Llamatepec volcano struck the Western region of El Salvador. In consortium with ICCO (an inter-church organisation for development cooperation) of Holland, Lutheran World Federation and Catholic Relief Service, Oxfam America designed a project with Salvadoran NGOs working in this region.

The PRVAS project was implemented in five municipalities in the departments of Sonsonate and Ahuachapán. This part of El Salvador suffers periodic mudslides, volcanic eruption, damaging winds and flooding and is dominated by the monoculture of coffee. Many poor communities are located inside expansive coffee farms (*fincas*), where families work in coffee production with limited access to services or ownership of land and earning little if any salary. These communities depend on landowners for their survival in a feudalistic model of social organisation. The three areas of work were as follows.

1 Forming a consortium and unifying methodologies

It was critical to bring the various NGOs working in the region together to agree to focus on the same basic elements of DRR. Since DRR had come to the attention of NGOs, they had developed different approaches for how to address it and to integrate it into their areas of expertise (e.g. health, environment, agriculture, etc.). These approaches brought diversity to the way DRR was being tackled, but also generated confusion and duplication at the local level.

To get NGOs to agree on shared elements of a DRR framework, the consortium members facilitated training, experience sharing and support. The interaction between Caritas, which focuses on community organising, development and agriculture, and PRO-VIDA, which specialises in water and health, provides a good example. When Caritas wanted to train the communities it worked in on water and sanitation in emergencies, the coordinator from PRO-VIDA was invited to give the workshop. The monthly meetings and trainings were important for team building and strengthening the common commitment to the project goals among the five NGOs. This collaboration and spirit of working together in the first year was the basis for the formation of networks between various communities throughout the region and the various NGOs and government actors later in the project.

2 Community organising and increasing capacity/knowledge

This component was the heart of the project. It required the basics of organising the communities (many of whom were hesitant to organise), forming community-level commissions, giving ongoing training and learning opportunities, while encouraging participation in forums, exchanges and advocacy efforts.

The first step was to inform communities about the existence of the Law for Civil Protection. The second step was educating communities about the Law itself and what their rights under it were. The progress of every community and municipality differed, but required consistency, dedication and time.

Residents analysed their vulnerabilities and capacities, undertook current risk mapping, mapping of past disasters, created early warning systems and evacuation plans and formed local committees for civil protection (CCPC). These were legally recognised by the municipal civil protection representatives. In years three and four of the project, the communities formed networks, linking individual communities and their community-level commissions with others throughout the municipality. One or two elected leaders of the network were appointed to be representatives in the municipal commissions for civil protection (CMPC).

3 Advocacy and relationship building with authorities

Community organising was a pre-requisite for the advocacy actions and relationship building with authorities. In the first year of the project, these activities were minimal due to fears and discomfort in dealing with authorities. As is common throughout the country, Ahuachapán and Sonsonate have a very polarised political environment, with political views (stemming back to the civil war during the 1980s) staunchly divided into left and right. The project worked with municipal leaders regardless of political persuasion and effectively overcame traditional party divisions by focusing on the provisions set out by the law and depoliticising the topic of DRR. Likewise, certain municipal authorities were less inclined to take a lead on the formation of their commissions (which according to the law is their responsibility), in which case the project opted to work at the community level, to pressure the municipality to form its commission.

As the project developed and matured, communities gained the confidence to interact with municipal actors and to voice their own needs. This was fostered by communities initiating small disaster-mitigation projects such as building a pedestrian bridge over a dangerous river crossing. This was a collaborative effort between communities (they provided the labour), PRVAS (seed funding) and authorities (funding and machinery). Residents have reported to Oxfam that now municipal authorities take their needs more seriously. They attribute this to two reasons. First, the needs they are expressing are based on an analysis of the situation, which shows the need has a basis identified and prioritised by the community and is not just another petition for infrastructure to come out of thinly stretched municipal budgets. Second, since the communities are linked into an entire network, their requests and pressure are regarded as coming from all of the communities in their municipal network.

Based on this model, in 2008 two other vulnerability reduction projects were created in other areas of the country. The first project called, PRVAMSS (Reducing Vulnerabilities in the San Salvador Metropolitan Area), was adapted to urban realities. Here, residents were better organised but the project faced challenges around population density, lack of available land and other issues prioritised by local authorities. The second rural project was designed for the mountainous regions of the department of Usulután, in the municipalities of Berlin and Algeria. The decision to expand to other regions of the country was in recognition of many of the same components in communities around the country, including the dispersion of many NGOs working on the topic without unifying their efforts, as well as a need for investment in DRR due to the high levels of vulnerability in those regions.

By forming a regional consortium of NGOs and donor agencies, the duplication of work has been avoided, allowing for better sharing of experiences and more effective leveraging of scarce funds for DRR. Likewise, by linking communities to local government, both directly and through NGOs, the processes of dialogue and advocacy becomes more sustainable. Communities realise they have the right and ability to continue that dialogue and to work to decrease their vulnerability after project funds end.

Figure 11.4 summarises the relationships that were built in the course of this work (Oxfam 2011).

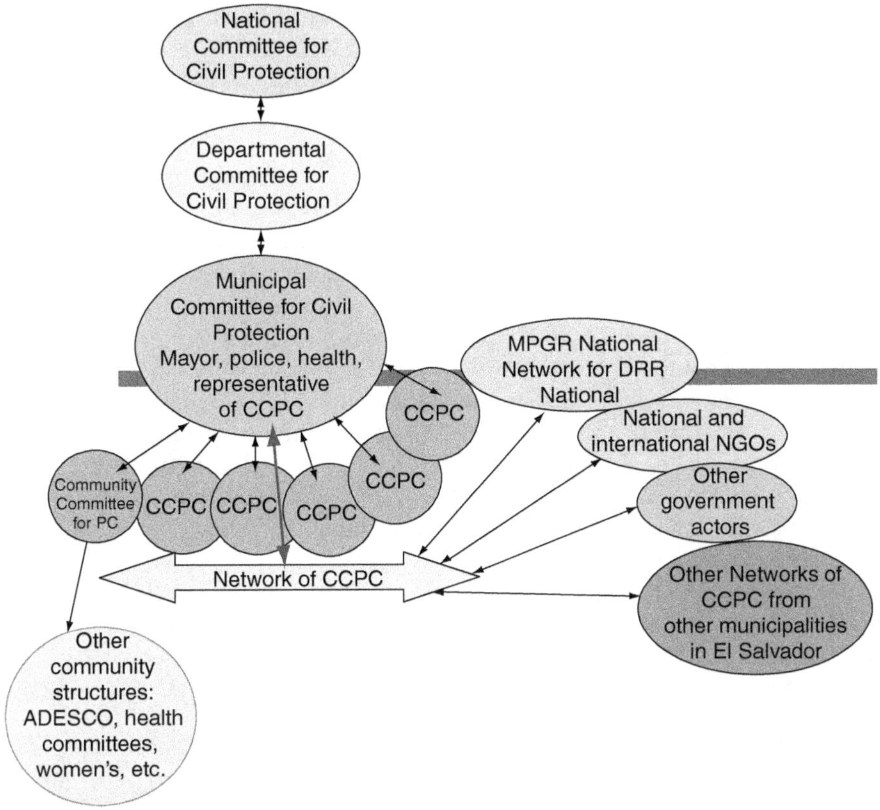

Figure 11.4 Advocacy and relationship building with authorities.

Oxfam America also observed that the approach of community organising and network creation allows for more rapid and effective evacuation during emergency, successful advocacy to local government for the construction of mitigation projects and increased awareness at the local, municipal and national level about the vulnerability of marginalised rural and urban communities.

For example, in 2010, during tropical storms Agatha and Alex, the community-level civil protection commissions responded by gathering data on damage and needs in their communities and coordinated with the municipal-level civil protection commissions for immediate response, resulting in authorities taking a more proactive role in these communities, attending to their immediate post-emergency needs. In the words of a local leader, before this project, when an emergency happened, the communities would 'wait to see' if any municipal authorities would show up to see if there were damages. The local communities have become advocates for their own needs and rights.

Urban CBA/CBDRR in Indonesia (Mercy Corps 2011)

The Asian Cities Climate Change Resilience Network (ACCCRN) is a project supported by the Rockefeller Foundation (2009–2013), in which the integration of DRR and CBA was demonstrated at the local level. ACCCRN is focused on participatory design, planning and budgeting of actions to build resilience to climate-related hazards. DRR is the entry point used to introduce climate change impacts and urban adaptation to the community. At the city level, climate change impact and adaptation were introduced to stakeholders through group dialogues, scientific vulnerability assessment and sector studies. Representatives from city government, local universities and local NGOs were involved in identifying urban vulnerability to climate-related disasters and potential impacts.

Semarang, one of ACCCRN's pilot cities in Indonesia, has been affected by extreme climate events. Located in Central Java, Semarang is Indonesia's fifth largest city with a total population of about 1.5 million people living in an area of 373 km^2. The city faces challenges that are exacerbated by climate changes: flooding, water scarcity and salt water intrusion, drought, water and vector-borne diseases and landslide. Flooding alone is responsible for almost 70 per cent of all hazards in the city. Assuming sea-level rise of 0.8 m over the next 100 years, Semarang City would lose over 8,000 ha in zones at distances ranging from 1.7–3.0 km inland (ISET 2011).

Integration of CBDRR and CBA at the local level was conducted through a pilot project in Tandang sub-district in Semarang. Tandang has a population of 8,407 inhabitants or 2,605 households, of which 14 per cent are female-headed. It is a poor community perched on steep hillsides that was created by a government relocation programme years back.

The pilot project undertaken at this level was documented and collaborated with the other three pilots in Semarang sub districts for upscaling as part of wider city resilience actions. Diponegoro University, through their Centre of Planning and Public Participation, co-facilitated this pilot in Tandang sub-district and started by selecting the most vulnerable neighbourhood units which were frequently exposed to natural hazards, such as landslide and strong winds, in which the latter posed a significant threat. The pilot intended to stimulate cooperation between a local university (that facilitated this project in this area), local community (particularly the most vulnerable groups) and local government, to diagnose the existing hazard and coping activities during extreme events based on community perspective and how this initiative could improve their capacity. The community established a Local Disaster Preparedness Committee, developed a Local Action Plan for the Tandang community, selected vegetation for landslide prevention and developed an early warning system.

The four-month pilot project utilised community participatory processes as their main approach, through holding a series of focus group discussions (FGDs) throughout the period of the project. The first three FGDs were aimed to enhance the community's capacity by sharing information about likely climate change impacts on the immediate surroundings. These group discussions also identified current coping activities. Based on outcomes from FGDs, the community was able to identify its own priorities: improving micro-drainage systems, reforestation, solid waste management and institutional capacity improvement.

The Local Disaster Preparedness Committee (LDPC) focused on ensuring the sustainability of the disaster preparedness system, especially in the event of landslide and cyclone. The intention was that various LDPCs prepare contingency plans and options for early warning systems, and these would be addressed within a city-wide resilience strategy – the work of a City Working Group.

According to Indonesian development planning regulations, there is a participatory planning mechanism in place that allows community action plans to be funded through local government budget requests which are part of the annual national government budgeting process. One of the challenges was to find ways to improve stakeholders' capacity at the city level to better advocate in the budgeting process for CBDRR and CCA initiatives. The creation of a multi-stakeholder committee at the city level – consisting of institutions working on DRR and climate change. At the beginning of ACCCRN project promoted ways to improve coordination among government agencies, local NGOs and universities in managing these issues. Figure 11.5 (Mercy Corps 2011) sets out this process of building city resilience.

Potential for upscaling

What can be seen from the foregoing is that a comprehensive approach to CBA/CBDRR has great potential for becoming an integral part of a 'no regrets' national/international development and climate adaptation strategy. Hopefully, this obvious direction will be taken up and reinforced at Rio + 20 (2012) and in the next iteration of work on the Millennium Development Goals (post-2015) as well as in whatever strategy follows the Hyogo Framework of Action, the UN's non-binding recommendations for DRR that expire in 2015.

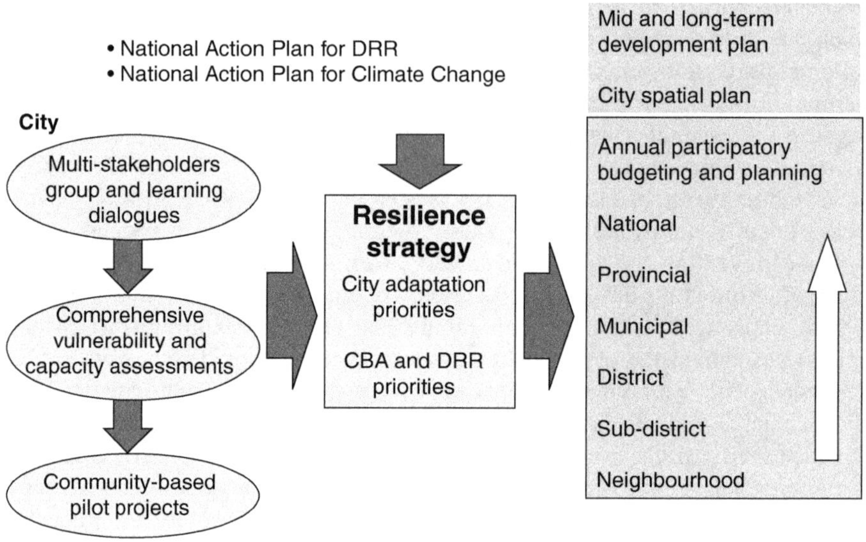

Figure 11.5 The process of building city resilience.

The linking up of communities into networks in the two case studies suggests that the well-established 'top-down' science of watershed management could develop its indigenous/autonomous mirror 'from the bottom-up.' As noted earlier, this kind of networking across communities has not been common in CBDRR work in the past. A language and institutional/academic basis for watershed management exists and could provide an entry point in meetings such as Rio + 20 for discussions that valorise local community knowledge and action within networks. Quantum leaps in development of hybrid knowledge for reducing uncertainty along these lines are certainly possible, but they must be based on respect and trust between those with local knowledge and outside specialists and a dialogue between them.

The connection with mainstream development is also not far to seek. Both CBDRR and CBA share a focus on local livelihoods. Also, the practical experience of CBDRR and CBA in the field shows that low-income communities buy into and sustain an approach that simultaneously provides jobs and income. This also implies that one should consider not just the hazards of climate change but also local and regional opportunities it might create. A common livelihood-based resilience approach could be mainstreamed as nothing less than the post-2015 paradigm for human development.

Conclusions and recommendations

We believe there is a compelling case for the implementation of a comprehensive and integrated approach to CBA and CBDRR that focuses on livelihoods and attempts to overcome the limitations of both separate approaches. The former is stronger in appreciation of the physical side of risk (the climate-related hazards), site, situation and exposure, but is weaker on the social aspects of risk. The latter is strong on the social aspects but often deficient when it comes to ecology, topography, the magnitude, frequency and areal extent of hazards (Miller *et al.* 2010).

We agree with Schipper and Pelling (2006) that '[t]he climate change community is … in danger of wasting time and money re-inventing the wheel if this knowledge is not soon translated from disaster risk reduction into climate change adaptation' (see also O'Brien *et al.*, 2006). However, it also true to say that the DRR community and practitioners of CBDRR need to take on board the more detailed engagement with the natural and physical world that characterises climate science and CCA generally, and CBA in particular.

Neither CBA or CBDRR take a truly long-term perspective into account, despite the former's nominal concern with very long-term, generational changes such as sea-level rise. An intergenerational approach to livelihoods and the resources required to sustain them would benefit both – harking back to the Brundtland Commission's classic discussion of intergenerational solidarity (WCED 1987).

However, integrating CBA and CBDRR in this manner will not be easy. Obstacles must be overcome. The following are recommendations that address some of the main obstacles (Pelling and Schipper 2009; Puhlin *et al.* 2010).

In the Asian context Puhlin and colleagues recommend a series of actions to facilitate an integrated approach to DRR and CCA. These are in the areas of information and communication, assessment methods, education and research and funding (Puhlin *et al.* 2010: 453ff.). These address routine problems that beset all development initiatives. Given political will and funding, the rest of these could fall into place; however, another problem would continue to block top-down initiatives even if funding were available and education, communication strategy and assessment methods were harmonised.

Current approaches to CBA and CBDRR are fragmented across different programmes with a high degree of silo working and institutional duplication leading to very low returns on investment for resource providers at the impact/outcome level. Therefore *programming for both CBA and CBDRR needs to be coordinated by a single governmental body, ideally one with the mandate and power to broker cooperation among ministries, departments, the private sector and civil society.*

Even then, impact of such action at the local level is likely to be minimal unless people have access to the resources needed to adapt in ways suggested by CBA/CBDRR exercises. The resource question is key, and therefore another of Puhlin *et al.*'s (2010: 464–465) recommendations must be underscored: *ensure that people, especially the poor, have access to natural resources and the environmental services that flow from them* (cf. Sanderson 2012). They must have the spaces to voice their interests and participate in meaningful ways in determining their development pathways.

References

Amin, S. 2011. *Maldevelopment: Anatomy of a Global Failure*, 2nd edn, London: Pambazuka Press.

Anderson, M. and Woodrow, P. 1998. *Rising from the Ashes: Development Strategies in Times of Disaster*, Boulder, CO: Lynne Rienner.

Birkmann, J. and von Teichmann, K. 2010. Integrating Disaster Risk Reduction and Climate Change Adaptation: Key Challenges – Scales, Knowledge, and Norms, *Sustainability Science*, 5, 2: 71–184.

Cuny, F. 1983. *Disaster and Development*, New York: Oxford University Press.

Delica-Willison, Z. and Gaillard, J.C. 2012. Community action and disaster, in: Wisner, B., J.C. Gaillard and I. Kelman, eds, *The Routledge Handbook of Hazards and Disaster Risk Reduction*, pp. 711–722, London: Routledge.

GNDR (Global Network of Civil Society Organisations for Disaster Reduction). 2011. *Unless We Hold Hands: Views from the Frontline 2011*, London: GNDR.

Hewitt, K., ed. 1983. *Interpretations of Calamity*, Boston, MA: Little and Brown.

IPCC. 2012. Summary for Policymakers, in Field, C.B., V. Barros, T.F. Stocker, D. Qin, D.J. Dokken, K.L. Ebi, M.D. Mastrandrea, K.J. Mach, G.-K. Plattner, S.K. Allen, M. Tignor and P.M. Midgley, eds, *Managing the Risks of Extreme Events and Disasters to Advance Climate Change Adaptation* [A Special Report of Working Groups I and II of the Intergovernmental Panel on Climate Change], pp. 1–19, Cambridge and New York: Cambridge University Press.

ISET. 2011. Semarang Climate Change Resilience Strategy, Indonesia. www.acccrn.org/sites/default/files/documents/6_Semarang_Resilience_Strategy_0.pdf (accessed December 2012).

Knowles, S. 2011. *The Disaster Experts*, Philadelphia, PA: University of Pennsylvania Press.

Maskrey, A. 1987. *Community Based Disaster Management*, Oxford: Oxfam.

Mercy Corps. 2011. *Building Synergies of DRR & CBA in the Urban Context – Learning from ACCCRN Indonesia Project*. Presented on the 5th International Conference Community Based Adaptation, Dhaka.

Miller, F., Osbahr,, H., Boyd, E., Thomalla, F., Bharwani, S., Ziervogel, G., Walker, B., Birkmann, J., Van der Leeuw, S., Rockström, J., Hinkel, J., Downing, T., Folke, C. and Nelson, D. 2010. Resilience and Vulnerability: Complementary or conflicting concepts? *Ecology and Society* 15, 3: 11. www.ecologyandsociety.org/vol. 15/iss3/art11 (accessed 15 March 2012).

Mitchell, T., Ibrahim, M., Harris, K., Hedger, M., Polack, E., Ahmed, A. and Hall, N. 2010. *Climate Smart Disaster Risk Management*, Brighton: IDS.

O'Brien, G., O'Keefe, P., Rose, J. and Wisner, B. 2006. Climate Change and Disaster Management. *Disasters* 30, 1: 64–80.

Oxfam. 2011. Paper presented on Oxfam America work at the 5th International Conference Community-Based Adaptation to Climate Change: Scaling Up Beyond Pilots, 24–31 March, Dhaka.

Oxley, M., 2011. Paper presented at the 5th International Conference Community-Based Adaptation to Climate Change: Scaling Up Beyond Pilots, 24–31 March, Dhaka.

Pelling, M. and Schipper, L. 2009. Climate Adaptation as Risk Management: Limits and Lessons from Disaster, *IHDP Update*, 2: 29–34, International Human Dimensions Programme on Global Environmental Change (IHDP) www.ihdp.unu.edu/article/read/managing-extremes-learning-from-past-experience (accessed 15 March 2012).

Perrow, C. 2007. *The Next Catastrophe*, Princeton, NJ: Princeton University Press.

Practical Action and Duryog Nivaran. 2011. *The South Asian Disaster Report*. Colombo, Sri Lanka: Practical Action.

Pulhin, J., Shaw, R. and Perreira, J. 2010. Disaster Risk Reduction and Climate Change Adaptation: Key Challenges and Ways Forward, in Shaw, R., J. Pulhin and J. Pereira, eds, *Climate Change Adaptation and Disaster Risk Reduction: An Asian Perspective*, pp. 451–476, Bingley: Emerald.

Sanderson, D. 2012. Livelihood Protection and Support for Disaster, in Wisner, B., J.C. Gaillard and I. Kelman, eds, *The Routledge Handbook of Hazards and Disaster Risk Reduction*, pp. 697–710, London: Routledge.

Schipper, L. and Pelling, M. 2006. Disaster Risk, Climate Change and International Development: Scope and Challenges for Integration, *Disasters* 30, 1: 19–38.

Shaw, R., Pulhin, J. and Pereira, J. 2010. Preface, in Shaw, R., J. Pulhin and J. Pereira, eds, *Climate Change Adaptation and Disaster Risk Reduction: An Asian Perspective*, pp. xix–xx, Bingley: Emerald.

Thompson, M. 2012. Civil Society and Disaster, in Wisner, B., J.C. Gaillard and I. Kelman, eds, *The Routledge Handbook of Hazards and Disaster Risk Reduction*, pp. 723–736, London: Routledge.

UNEP (United Nations Environment Programme). 2009. *Reducing Risk Through Environment in Recovery Operations*, Nairobi and Geneva: UNEP & UNISDR.

UNISDR (United Nations International Strategy for Disaster Reduction). 2011. *Global Assessment of Disaster Reduction 2011*, Geneva: UNISDR.

Wangui, E., Smucker, T., Wisner, B., Lovell, E., Mascarenhas, A., Solomon, M., Weiner, D., Munna, A., Sinha, G., Bwenge, C., Meena, H. and Munishi, P. 2012. Integrated Development, Risk Management and Community-Based Climate Change Adaptation in a Mountain-Plains System in Northern Tanzania. *Journal of Alpine Research*, 100, 1, www.rga.revues.org/pdf/1701 (accessed 15 March 2012).

WCED (World Commission on Environment and Development). 1987. *Our Common Future*, New York: Oxford University Press.

Wiggins, S. 2011. *Adaptation United*, London: Tearfund.

Wisner, B. 2012. Violent Conflict, Natural Hazards and Disaster, in Wisner, B., J.C. Gaillard and I. Kelman, eds, *The Routledge Handbook of Hazards and Disaster Risk Reduction*, pp. 71–82, London: Routledge.

Wisner, B., Blaikie, P., Cannon, T. and Davis, I. 2004. *At Risk*, 2nd edn, London: Routledge.

Wisner, B., Mascarehas, A., Bwenge, C., Smucker, T., Wargui, E., Weiner, D. and Munishi, P. 2012a. "Download Let Them Eat (Maize) Cake: Climate Change Discourse, Misinformation and Land Grabbing in Tanzania" *Let Them Eat (Maize) Cake: Climate Change Discourse, Misinformation and Land Grabbing in Tanzania*, www.cornell-landproject.org/download/landgrab2012papers/wisner.pdf. "*Contested Global Landscapes*" Contested Global Landscapes. A Multidisciplinary Initiative of the Cornell Institute for the Social Sciences, www.cornell-landproject.org/. Ithaca, NY: Cornell University.

Wisner, B., Gaillard, J.C. and Kelman, I. 2012b. Framing Disaster: Theories and Stories Seeking to Understand Hazards, Vulnerability and Risk, in Wisner, B., J.C. Gaillard and I. Kelman, eds, *The Routledge Handbook of Hazards and Disaster Risk Reduction*, pp. 18–34, London: Routledge.

World Bank. 2011. *Rising Global Interest in Farmland: Can It Yield Sustainable and Equitable Benefits?* Washington, DC: World Bank.

12 Upscaling CBA in agriculture

Stephen Baas and Anna Ricoy with contributions from S. Ramasamy, C. Scaramella, M. Sta Catalina and C. Zanev

Agriculture is the main source of livelihood for an estimated 86 per cent of the three billion rural people in developing countries (World Bank 2008). About 70 per cent of the world's poor people reside in rural areas and live on less than $1.25 per day (IFAD 2010). Agriculture plays a critical role in efforts to reduce hunger and malnutrition and to improve the living standards of the rural poor, including smallholders and landless workers.

Despite large increases in global agricultural productivity during the past decades, 925 million people are currently living with chronic hunger and malnutrition (FAO 2010). To meet the food demands of future populations, projected to increase from the present 6.7 billion to 9.1 billion by 2050, requires an increase in global food production of 70 per cent (FAO 2009). This increase has to be achieved in times when the world is facing a range of additional global challenges for food security, including: the impacts of extreme events and climate change, increasing scarcity of natural resources, rising inequality and problems of access to food by the most vulnerable populations and increasing food price volatility as a result of climate change, misplaced policy responses to it and many other factors (WFP 2011a; Parry *et al.* 2009).

Adaptation in agriculture is crucially important to feed a growing population and because the majority of the rural population in the developing countries depends on agriculture for their livelihoods. Adaptive capacities need to be enhanced at all levels in a systematic and coherent way with a particular focus on the poor and vulnerable. Durable solutions require integrated approaches that include increasing agricultural production and productivity while taking care of the environment, improving access to nutritious food, and enhancing the risk-coping capacities, particularly of the poor, to better withstand current and future risks and shocks. Adaptation is hence not a single intervention; it is a process incorporating initiatives that range from addressing the underlying drivers of vulnerability to responding to the impacts of climate change. It also requires action at multiple levels: community-based adaptation (CBA) provides a crucial pathway to enhance capacities for adaptation at local levels; it is equally important, though, to put supportive institutional mechanisms, technical capacities and policies in place.

A wide range of projects and success cases of CBA exist, as illustrated at the international CBA Conferences 2008–2011. However, a strong need remains for wider upscaling of CBA, including through key sectors in order to enhance outreach and build capacities for addressing the challenges of climate change. Upscaling is typically presented as a concept covering all forms of expansion, growth and replication (Hooper *et al.* 2002). Upscaling implies different dimensions: (1) *horizontal upscaling* – making local lessons apply to wider spatial scales (in terms of outreach); (2) *vertical upscaling* – creating a supportive policy and institutional environment to enhance successful initiatives from local to national/regional/global levels; (3) *temporal upscaling* – making efforts sustainable over time (and encompassing both the horizontal and vertical dimensions). For upscaling CBA in agriculture these definitions imply wider replication of location-specific processes and good practices for CBA in other places with similar agro-ecological contexts (geographic outreach). It also means empowering communities and local institutions working with farmers and strengthening capacities of district/national governments, NGOs and private sector organisations to support CBA processes and remove constraints for adaptation.

This chapter consolidates lessons from a range of field projects implemented by FAO and WFP with government partner institutions in several regions of the world. It discusses the relevance, role and potential of the agriculture sector in upscaling CBA. The first section looks at the context required to enable CBA among vulnerable farmers and communities and discusses entry points for CBA in agriculture. The second section addresses the scope for upscaling CBA in agriculture and explores conditions and options for doing it successfully.

Enabling CBA

The context for CBA in agriculture

Agriculture-based livelihoods are highly sensitive to the impacts of climate variations and change (FAO 2011a). This is due to the high exposure and often limited adaptive capacities and resilience among rural communities, particularly in developing countries, who depend on agriculture for their food security (GIZ 2011). The combination of both aggravates production risks for farmers, livestock keepers, fishers and forest-dependent people, particularly in regions that already suffer from chronic soil and water scarcity and frequent exposure to natural hazards (FAO 2008a). Communities in fragile environments, such as dry lands, mountain areas and coastal zones, especially poor and smallholder farmers, women and marginalised groups such as indigenous people, are most vulnerable (FAO 2008a). Farmers in developing countries report increasingly about their observation that agricultural seasons are shifting, rainfall is getting more unpredictable and often more erratic and temperatures are rising, with more drought spells than in the past (FAO project reports, 2008–2011; Macchi *et al.*, 2011). Climate change also leads to changes in the frequency, intensity, spatial extent, duration and timing of extreme weather and climate events (IPCC 2012). This has a range of impacts on the food and nutrition security of poor rural communities beyond production itself, including food price increases that limit access to nutritious food. Declining agricultural production also affects employment and income opportunities, further decreasing poor people's purchasing power and limiting their abilities to invest in assets that would contribute to building adaptive capacity (WFP 2011a; Parry *et al.* 2009). Climate-related shocks force the many poor farmers without access to insurance, safety nets or other risk-management tools and services to resort to negative coping mechanisms or 'maladaptation', which further undermine their livelihoods and trap them in poverty and food insecurity (WFP 2011b).

CBA to climate change is a community-led process, based on communities' priorities, needs, knowledge and capacities, which should empower people to plan for and cope with the impacts of climate change (IIED 2009). The adaptive capacity of the communities is determined by a broad range of factors, including their access to economic, social and institutional resources (Füssel 2007). Local institutions are the fundamental drivers of CBA processes since they shape the effects of climate hazards on communities in three important ways: (1) they influence how households are affected by climate impacts; (2) they shape the ability of households to be prepared for and respond to climate variability and hazard impacts and how they pursue different adaptation practices; and (3) they mediate the flow of external interventions in the context of adaptation (Agrawal *et al.* 2008). But communities and farmers also depend on a variety of social, political and economic institutions at national and global levels. Effective CBA therefore needs to link and integrate measures originating from communities and their members (autonomous adaptation) and policies or strategies promoted from the government side (planned adaptation) to facilitate adaptation to climate change and reduce the risk of maladaptation (Figure 12.1). The two types of adaptation are always interlinked and both are required (Ayers and Anderson 2011). FAO emphasises the operational space that CBA creates through the synergies between bottom-up adaptation initiatives and top-down interventions at different scales (FAO 2008b).

The spectrum of CBA actions reflects local situations and needs, as well as the diversity of interactions between communities, governments and other institutions. Local actions for CBA mirror the complexity of people's livelihoods; they are not limited to sector boundaries. There is in fact no one-size-fits-all solution or uniform approach to CBA. CBA is a constant learning process that gradually builds and shapes the capacities needed to cope with climate change-related impacts. These processes can be highly diverse since they take place in a broad range of different ecosystems, socio-economic settings and vulnerability contexts (FAO 2011b). However, agricultural and natural resource management practices and processes that reduce vulnerability to climate risks and enhance people's livelihood resilience are key aspects of CBA in almost all rural communities in developing countries (FAO 2008b; Macchi *et al.* 2011).

To make community-based approaches to adaptation in agriculture effective, local capacities and institutional mechanisms need to be set in place and empowered to identify, test, validate and monitor traditional or newly introduced

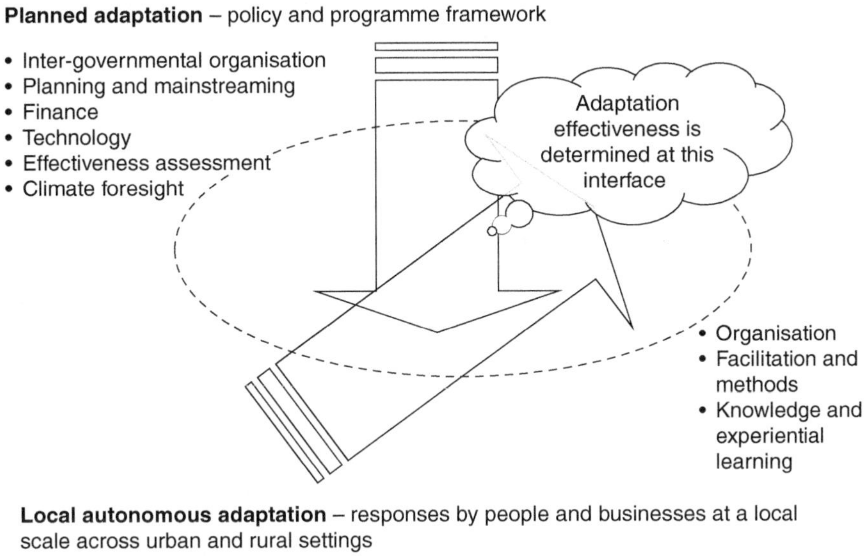

Figure 12.1 CBA at a nexus between autonomous and planned adaptation (source: Ayers and Anderson 2011).

agricultural and natural resource management practices and processes that reduce vulnerability to climate risks and enhance people's livelihood resilience.

Entry points for CBA in agriculture[1]

Building on existing knowledge and practices

Coping with climate-related risks is not new for rural communities. For centuries, farmers and herders have experienced climate variability and extreme weather events and adjusted their crop and livestock management accordingly. At this point, however, these adjustments require different scales and dimensions as climate change adds additional stresses in terms of speed and intensities. Local communities need to consider bigger and quicker changes, with higher variability to occur than ever before, and more difficult to predict.

CBA in agriculture can build on a large pool of existing local and scientific knowledge. FAO experience (project reports 2008–2011) suggests that the value of local and traditional agricultural practices is essential to shape long-term adaptation. FAO projects have led to successful results when replicating indigenous and known local practices in areas where they were not practised before, and by complementing them with other practices based on scientific knowledge. This included, for instance, the integrated testing of indigenous and research-based salt and submerge-tolerant rice varieties in Bangladesh (BRRI-FAO 2009), and of fruit trees tolerant to drought spells in the Cordillera Region of the

Philippines (Department of Agriculture of Cordillera Region of Philippines 2012), and their transfer into other, new locations with similar agro-ecological characteristics, where they have proven to be also environmentally, economically and socially beneficial.[2]

Upgrading and climate proofing of small-scale agricultural and community infrastructures such as hazard-resistant animal shelters, grain reserves or small-scale irrigation, drainage, water ponds, small dams or reservoirs is another entry point for CBA in agriculture. The documentation on joint projects with FAO provided by the Centre for International Agricultural Development of China (2009), the agricultural extension services of the Ministry of Agriculture, Food Production, Fisheries and Rural Development, Saint Lucia, (2011), the Ministry of Agriculture, Grenada (2008) and the Ministère de l'Agriculture, des Ressources Naturelles et du Développement Rural, République d'Haïti (2007) provide a range of location-specific good practices. Another example with wide outreach is the establishment of community granaries by rural women, which was supported by WFP in drought-stricken areas of Cameroon, reaching 750,000 beneficiaries. Initial cereal stock and management training was provided to women's groups to set up and manage community granaries under a food-for-work programme. The mobilisation of women around the granary activity was an important channel to undertake other resilience-building activities relating to health and nutrition, agro-forestry and sustainable land management and agriculture. After one year of operation of the granaries, the number of hectares cultivated and the level of food production had increased significantly; no emergency operations have been necessary in the region since the setting up of the community granaries (Urquhart 2010).

Sustainable natural resources management (SNRM) is an equally important pillar for CBA in agriculture to build on. SNRM can provide communities with multiple benefits such as reduced erosion, increased soil fertility, improved water quality and more resilience against climate extremes, if done properly. CBA approaches which include SNRM are likely to reduce future risks and hazard impacts to a significant extent. One illustrative example is Ethiopia's MERET (Managing Environmental Resources to Enable Transition) Programme, implemented through the Natural Resource Department's extension system and supported by WFP, which enables food-insecure communities to manage their natural resources effectively in order to increase their resilience to weather-related shocks. Under MERET, a community-selected management committee works with extension agents to identify their priorities, select and plan SNRM activities. Most of these activities not only reduce risks, but also deliver direct agricultural and food security benefits. For example, leguminous fodder plants combine gully stabilisation with fodder supply for livestock. Similarly, pigeon pea provides a source of nutrition, stabilises land and is drought-resistant (Nedessa and Wickrema 2010). There are numerous other examples and a wealth of knowledge on existing SNRM technologies and practices, which are suitable to address current and future risks to a significant extent.[3]

Land use planning has the potential to address current and expected future risks, including possible shifts in cropping and grazing patterns, farming systems and ecosystems. For example, hazard-specific local vulnerability maps featuring location-specific overlays between agriculture land, settlements and current and future climate risk patterns can result in land use recommendations to avoid land degradation and risk exposure. The joint FAO/government of Hungary project conducted over 2007–2009 in the Bereg Region of the upper Tisza river provided a good example of how participatory local territorial planning with 14 communities led to the re-establishment of a polder-based water retention scheme to buffer regular seasonal flooding in the Bereg and downstream. This included the planning of land use change from cropping to integrated grazing (livestock production) and forestry (FAO Project summary report 2007, unpublished). Overlaying climate risks with food security data allows for targeting the communities most vulnerable to climate-related food insecurity and can also guide CBA planning. Similarly, linking food security, vulnerability and livelihood analysis with climate information provides an important entry point for guiding CBA in food insecure countries (see WFP 2011a and below).

Linking short- and long-term perspectives

Field experience shows that the process of shaping adaptation, particularly at the local level, is well initiated by addressing the current risks and vulnerabilities first, thus matching poor farmers' ways of planning (FAO 2011b). For poor and food insecure rural communities the reduction of current risks and vulnerabilities is of higher priority than addressing long-term adaptation needs. Farmers are in principle flexible to adjust and change crop varieties and production patterns every season. The majority of smallholder farmers in developing countries apply a short-term perspective by thinking in terms of agricultural seasons rather than long-term climate change perspectives, and also because they are pressed by immediate needs and limited in their opportunities to make long-term investments. However, this short-term perspective provides an entry point for addressing the long term anticipated changes. It is crucial, particularly in the context of extreme poverty and acute food security crises, to simultaneously address immediate food and nutrition needs and longer-term adaptation challenges (WFP 2011a).

The Livelihood Adaptation to Climate Change project in Bangladesh, for example, focused on understanding past and current risks, vulnerabilities and associated impacts at the local level first, and then added expected long-term trends as a second layer of planning, reflecting the priorities of farmers and extension staff (FAO 2008b). Other FAO projects in Asia, Africa and the Caribbean have confirmed this experience. Important lessons learned are that CBA planning should be grounded on actions related to short-term climate variability and extreme events, and that effective climate risk management at the local level requires spatially and temporally differentiated climate information (FAO 2008b).

Building on existing institutional structures

CBA is a long-term process which needs to be flexible and shaped in a participatory and demand-responsive way. There is no need to create new organisations as drivers of CBA in agriculture; only where no suitable structures exist new ones may need to be established. CBA activities can be easily embedded into local realities within the context of existing activities and processes, such as farmer field schools, farmers' groups, cooperatives or associations. The development and transfer of climate change-related know-how and technology can be linked with other matters of concern to farmers, fishermen and pastoralists, such as integrated pest management, market information and access, veterinary services, micro-finance options, etc. so as to address the livelihood reality of local people in a holistic way.

A number of researchers, policy-makers and practitioners (Schipper and Pelling 2006; Mitchell and van Aalst 2008; Venton and Latrobe 2008; Gero *et al.* 2011) strongly advocate for the need to combine CBA and community-based disaster risk reduction (CBDRR) strategies and institutional mechanisms. With its often well-established structures and proven practices for reducing disaster risk at the community level, CBDRR is a good basis for CBA to build on. On the one hand, coping with the expected increase of extreme climate events expected from climate change can benefit from and build on proven community-level practices in reducing the risk of natural hazards such as drought, floods and cyclones. Building on the existing institutional structures for CBDRR also creates opportunities for better linking CBA from local to higher institutional levels. On the other hand, CBA will add to past and existing efforts in CBDRR by addressing the longer-term adjustments needed to adapt to changes in mean climatic conditions. As Mitchell *et al.* (2010) note, 'business-as-usual disaster risk management will fail without a significant shift in risk calculation and designing interventions with incorporation of climate modelling and associated uncertainty'. This requires additional know-how, which can be achieved in CBA- and CBDRR-related institutions through integrated capacity development.

Enhancing existing tools and methods to support CBA

There is a wide range of existing tools and methods in DRR and development, which, if adapted, can also serve in the context of climate change. For example, a wide spectrum of risk and vulnerability assessment methodologies for CBDRR exists, which can be fine-tuned and complemented by climate change impact modelling to address long-term climate change impacts as an additional challenge to the existing vulnerabilities (examples: Vulnerability Assessment Tool of the University of the Philippines Los Baños, 2011; CARE's Climate Vulnerability and Capacity Analysis Handbook, 2009). CBA can also use and build on early warning systems and seasonal climate and weather forecasts which are both part of short- and long-term climate risk management. WFP and FAO have developed a set of food security-specific early warning and risk management tools and services that can also support adaptation planning efforts at the local, national and regional levels, some of which are described in subsequent sections (WFP 2011a, 2011b; IPC Global Partners 2008; FAO 2007).

All the above entry points for CBA in agriculture call for long-term processes of participatory action research involving the local communities, other local institutions, NGOs, research and agriculture extension, the private sector and government. CBA in agriculture depends on strong community networks that bridge the information flow between individual farmers and agriculture support services. The farmer field schools (FFSs) approach is one of the proven mechanisms successfully introduced by FAO for decades that facilitated formation of farmers' networks. Continuous partnerships and mentoring of these groups by agricultural support services and research institutes is a fundamental feature of CBA in agriculture to catalyse two-way information flow and learning. It helps the farmers to better define adaptation practices according to their location-specific needs.

Upscaling CBA in agriculture

The agricultural sector is central in rural areas of the developing world and has the potential to act as an 'engine' for the integration and upscaling of CBA and CBDRR within agriculture and beyond. It has the capacity to mobilise, spread, replicate and aggregate grassroots initiatives and programmes to grow, be flexible and become sustainable. In many contexts, agricultural sector institutions are well placed to facilitate this integration and upscaling as they operate at all levels (national, provincial, district and local). Ultimately, however, we have to be aware that, whether adaptation in agriculture happens or not depends on the farmers' decisions and abilities to adapt, which in turn depends on a range of other factors supporting rural livelihoods, which may fall in the responsibility of other, non-agricultural institutions. Adaptation in agriculture, specifically local adaptation measures, cannot be prescribed, but it can be catalysed and enabled though appropriate incentives and enabling policies which may require cross-sectoral, multi-stakeholder institutional arrangements.

As stated in the introduction, upscaling is typically presented as a concept covering all forms of expansion, growth and replication, implying different dimensions, namely *horizontal*, *vertical* and *temporal.* Upscaling CBA and other similar approaches such as CBDRR in agriculture implies wider replication of location-specific good practices for CBA in other places with similar agro-ecological and socio-economic contexts. It also means empowering communities and local institutions working with farmers, and mobilising the support services of district/national governments, NGOs and private sector organisations to promote CBA processes. Thus, successful upscaling of CBA in agriculture requires a set of favourable conditions, combining bottom-up community-led processes with institutional and policy interventions shaped by national governments or international negotiations.

Creating favourable institutional and policy frameworks

Strengthening the coordinating capacity within agriculture sector agencies

Upscaling CBA in a sector like agriculture requires support of strong offices at all levels with clearly defined and coordinated responsibilities vis-à-vis CBA processes. Agriculture sector agencies (e.g. ministries of agriculture) have to coordinate the technical and institutional requirements needed to promote and upscale CBA in agriculture. They need to put operational networks and communication channels in place for the sector, which help to regularly assess sector-specific vulnerabilities and risks, analyse the livelihood requirements of farming communities and accordingly catalyse services for CBA from national to local levels and vice versa in a coherent and coordinated way.

Equally important is to ensure that the national coordinating agencies within the sector engage district-level agriculture departments, local operating extension officers, researchers and NGOs that contribute implementing agriculture-related CBA activities. Well-coordinated capacity-building processes for all kinds of service providers involved in CBDRR and CBA is a precondition for harmonising, delivering and sustainably upscaling CBA. At the local level, sector-specific support and guidance, particularly from agriculture as one of the most climate-affected sectors, is vital for any integrated CBA process.

There also remains a need in most countries for strengthening the representation of the agriculture sector institutions in existing national platforms and global negotiations for climate change adaptation (CCA) and DRR that provide national mechanisms for stakeholders' dialogue and coordination, as well as access to global funding mechanisms.

Multi-stakeholder involvement

Since farmers' livelihoods are not constructed along a single sector's objectives, CBA actions for agriculture must be aligned with other planning processes relating to the environment, water, nutrition, food security, as well as broader rural and urban development objectives, and hence require integrated, cross-sectoral planning and decision-making. Neither agriculture nor any other single sectoral intervention alone can provide sufficient measures to manage the future climate risks to food and nutrition security and agriculture. Many agriculture and food security risks that are accelerated by climate change, such as large-scale land degradation, far outstrip the ability of any single agency, let alone community, to resolve it (Nedessa and Wickrema 2010).

Multi-stakeholder engagement is essential as upscaling calls for expansion of existing participatory networks, alliances and partnerships to support CBA in an integrated way, going beyond agriculture. Wherever sector-specific technical options for integrated CBA processes are crucial, one sectoral line agency may take a lead in initiating and facilitating CBA processes, but at the same time other relevant sectors should provide support. Multi-stakeholder approaches also need to include local institutions to facilitate the ownership of the local communities involved.

Indeed, FAO experience indicates that cross-sectoral coordination and planning mechanisms are most likely to succeed in facilitating the dialogue with CBA institutions when they are initiated or replicated at the lowest administration levels. Recent experiences in Bicol Region in the Philippines provide a good-practice example: several agencies, under the leadership of the Department of Agriculture, signed a Letter of Agreement to ensure a multidisciplinary and multi-level approach to climate risk management including local government units. This case showed that cross-sectoral integration was easier when going all the way down to decentralised levels, i.e. local government, where different disciplines report primarily to the same community or municipality administration, and not only along hierarchical lines of sectoral agencies (see Box 12.1).

Another example of an integrated approach to achieving food and nutrition security among farming communities is a recently initiated five-year project under the Adaptation Fund in Ecuador. The project aims at strengthening the resilience and food security of 120 food insecure and vulnerable communities located in the Pichincha province and the Jubones river basin through community- and ecosystem-based adaptation. Here, the Ministry of Environment, which hosts the Secretariat for Climate Change, made institutional arrangements with the Ministry of Agriculture, Livestock, Aquaculture and Fisheries, local governments and WFP that allow for the implementation of activities at the watershed level, across communities and administrative borders (Government of Ecuador 2011).

Mainstreaming support for CBA into agricultural and rural development planning or CCA/DRR plans

Systematic CBA processes usually lead to local action plans for CBA. These community plans can be used systematically to inform sectoral development planning. The strategic integration and upscaling of CBA and CBDRR plans into regular agricultural and rural development planning, or into sector-specific CCA or DRR plans if they exist, is essential to ensure upscaling of CBA and the wider dissemination of good agricultural practices, technologies and local capacity-building processes for climate risk management and adaptation (Ministry of Agriculture and Fisheries, Belize, 2011). It can also support the mobilisation of resources (both technical support as well as financial) of line agencies to partner with CBA institutions to jointly implement local adaptation processes, as also illustrated by the example from Ecuador. In addition, mainstreaming CBA and CBDRR components into agriculture and rural development, CCA or DRR plans and programmes also allows for linking them to measures that address the structural factors of vulnerability, promotes empowerment of farmers and communities and allows them to gain access to the resources they need for successful CBA/CBDRR.

Box 12.1 Upscaling CBA in agriculture through institutionalisation of good practices for DRR and CCA in Bicol Region, Philippines

Bicol Region in the Philippines is exposed to high annual and seasonal rainfall variability and is highly vulnerable to natural disasters such as typhoons, flash floods, strong winds, thunderstorms and drought. Moreover, the seasonal variations in weather, especially of rainfall, cause variations in production and seasonal price fluctuations, which affect the income of the farmers significantly. It is expected that climate change will further exacerbate the region's exposure to climate-induced risks. This prompted the Philippine government to initiate a capacity development project with the Department of Agriculture (DA) Regional Field Unit in Bicol, and local government units (LGUs) to provide better assistance to poor and marginal farmers and fisherfolk in disaster-prone areas to reduce their vulnerability to climate variability and change, while maintaining or increasing their production levels. The project initiated a participatory action research process led by the DA, in collaboration with two Bicol universities, the Meteorological Agency (PAGASA) and selected municipalities, barangays, farmers' groups and farmers. It combined technical and institutional capacity-building measures for current DRR with medium- to long-term CCA perspectives.

Good practice options suitable to enhance DRR in agriculture, including in cropping, livestock and fisheries management, soil and water conservation, were identified and selected by a formalised group of DA experts, extension officers, researchers as well as LGUs and farmer representatives. The selected practices were then field tested with farmers' groups in selected barangays through an action

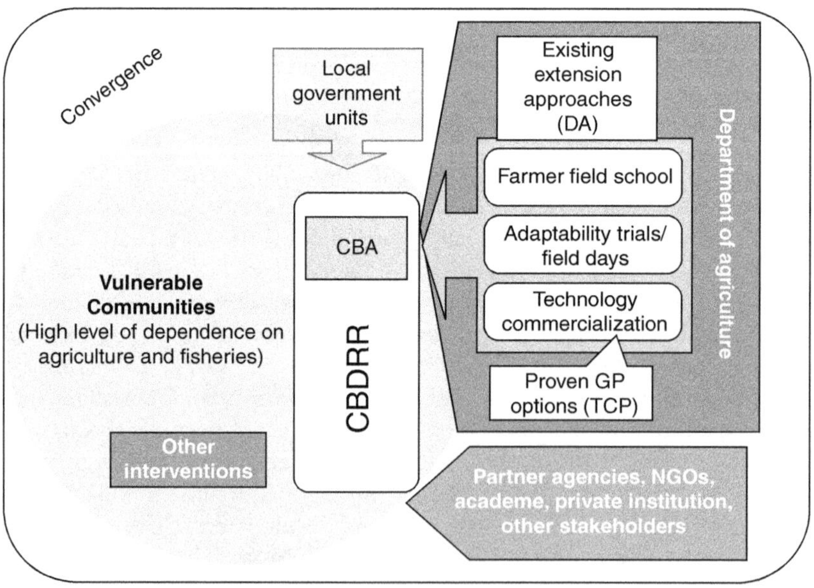

Figure 12.2 Participatory action research combining technical and institutional capacity building for DRR and CCA in Bicol, Philippines.

research approach facilitated by the agricultural universities and agricultural extension. Enhanced climate information products produced jointly by PAGASA and DA were designed for the specific needs of farmers and informed the season-specific selection of crop varieties for field demonstrations. Field days and exchange visits were organised between communities to share the learning. The project also initiated a stakeholders' dialogue for preparing a sector-specific Plan of Action for Enhanced DRR in Agriculture, owned by the DA.

To ensure replication and further upscaling of good practices and technologies for CBA/CBDRR, the participating agencies have initiated a joint Memorandum of Understanding to follow up and institutionalise within and among the agencies the continuation and further expansion of the successfully initiated joint action research process. Further replication of action research and successfully tested practices have been included into the annual development planning of DA (at regional level) as well as of the selected municipalities (LGUs), thus receiving funding for wider replication. Lessons learned and technical recommendations for replication beyond the selected communities/municipalities have been documented and shared with other regional and national stakeholders, and will be mainstreamed through the existing agricultural extensions services, Training of Trainers and FFSs.

See also: www.fao.org/climatechange/ph/bicol/en.

In Ethiopia the local-level participatory planning approach used in the WFP-supported MERET programme is reflected in the national 'Community-Based Participatory Watershed Development' guidelines, guiding the work of extension services (Ministry of Agriculture and Rural Development Ethiopia 2005). The impact of applying community-based approaches at scale can be seen, for example, in Ethiopia's Tigray region, where the Natural Resources Department applied the MERET approach to the entire region. For the past 20 years, about 80 per cent of cultivated land in Tigray has been treated with soil and water conservation measures. This amounts to about 960,000 ha, of which 300,000 ha are under livelihood-improving measures such as fruit trees, fodder shrubs and grasses (Nedessa and Wickrema 2010).

Providing services to promote CBA and guide upscaling

Enabling agricultural extension and research systems to promote technology transfer and dissemination

Agriculture extension systems are a main channel to identify and promote the dissemination of hazard-proof, climate-resilient and demand-driven technologies in agriculture. Climate change now calls for more, flexible action research to define what good practices are under different and changing hazard contexts, as well as diverse agro-ecological and socio-economic conditions in which farmers operate. Local adaptation processes in agriculture are best initiated through the promotion of known, local and/or indigenous practices, blended with complementary technologies provided by research to further widen the scope for resilience building. This includes adjusting planning for adaptation to agricultural planning cycles, and providing regular technical guidance from extension and research systems. The development of long-term options that represent viable choices for farmers is particularly important.

Upscaling CBA in agriculture requires the institutionalisation of these flexible action-research approaches. In order to capacitate existing agriculture extension and research systems to guide CBA action research processes in agriculture, renewed partnerships between researchers, academic experts, extension officers and farmers' representatives are needed. Expert groups can be set up under the leadership of agricultural departments to advise extension services (and other actors) on the selection of potential CBA good practices, and in evaluating the impacts and lessons from the action research. Once institutionalised, these technical expert groups can systematically advise the replication and dissemination of hazard-proof, climate-resilient and demand-driven technologies in agriculture. The setup of technical expert groups was tested in the Philippines and Bangladesh as a means to promote quality selection and upscaling of good practices. The groups catalysed the development of technically validated good practice option menus, which were also used by other actors in CBA, including NGOs, and thus contributed to a more consistent message about what good practices for CBA in agriculture are (DAE Bangladesh and FAO 2008; see also Box 12.1).

The private sector carries another important responsibility in the context of upscaling CBA related to technology dissemination, notably through fostering markets for new climate-resilient technologies, products and services in close consultation with the above-discussed extension and research systems. This includes the provision of more hazard-tolerant seed varieties, the promotion of enhanced seed systems, the creation of strategic arrangements of CBA farmers' organisations with seed suppliers and the promotion of direct marketing linkages.

Systematic knowledge management, communication and awareness-raising

Crucial for any local-level adaptation processes is awareness-creation and the understanding of the case-specific risks and opportunities which climate change induces at different levels, in order to identify, utilise and strengthen communities' adaptive capacities accordingly (GIZ 2011).

Extension and research processes for CBA imply an intensive use of knowledge and information, calling for appropriate communication, methods and tools to achieve stakeholders' awareness at all levels and support a continuous flow of learning. Harris *et al.* (2012) flag the need for partnerships that can vertically link knowledge and skills *between* scales – from the household to global – and horizontally promote shared learning through cross-disciplinary dialogue on a wide range of issues *within* scales.

Successful upscaling of CBA in agriculture largely depends on strong knowledge networks and the institutionalisation of demand-oriented agricultural information and communication services at national and regional level. At district and local levels clear communication strategies and simple messages are needed to strengthen the capacities of local stakeholders to engage in decision-making for risk reduction, and to ensure that local communities and farmers are directly involved in the design of CBA strategies and benefit from mutual learning. Lessons learned from FAO field projects reveal that in the context of CBA this can be achieved through the systematic design and use of participatory communication processes, strategies and media that link sound local knowledge with scientific knowledge and consider how rural services are delivered. It is crucial to ensure that all stakeholders in a particular agro-ecological context are enabled to participate in the knowledge sharing. Examples of successful communication strategies for CBA are the national communication campaign on climate-related disaster risk management in St Lucia (Jules 2011) and the cultural awareness-raising programme on CCA in agriculture implemented by the Livelihood Adaptation to Climate Change project in Bangladesh (including posters, leaflets, videos, folk and drama) (FAO 2008b). Both catalysed wider outreach and knowledge sharing beyond the farmers and community members directly involved in the CBA interventions.

Another specific example of enhanced knowledge management that improves the decision-making capacities of agricultural producers – now and in the future – is the provision of enhanced, tailor-made climate information at seasonal, monthly and ten-day timescales, transformed into agro-meteorological forecasting, and transmitted to farmers through agricultural extension. This climate information product, which was tested in the context of the Livelihood Adaptation to Climate Change project in Bangladesh, has good potential to facilitate the strategic decision-making of farmers and extension workers throughout the agricultural cropping cycles (ADPC and FAO 2010). Based on this experience the same approach was successfully replicated with PAGASA in Bicol Region of the Philippines in 2010/2011. These types of products, however, still need to be further technically upgraded and institutionalised.

Monitoring and evaluation as the basis for upscaling

As Ayers and Anderson (2011) note, despite the increase in attention to and practice of adaptation and the wealth of knowledge (including traditional, innovative, project and research experiences), there remains a lack of appropriate participatory, practical and replicable methodologies for measuring, monitoring and evaluating the results of adaptation in general, and CBA in particular. Having empirical evidence at the local level is, however, pre-conditional to support advocacy for upscaling CBA in agriculture.

Monitoring and evaluation (M&E) systems for CBA must be simple but nevertheless efficient to analyse the involvement of local actors in the effectiveness of CBA practices and processes. The sound understanding of what exactly makes certain livelihood groups particularly vulnerable or resilient to climate change informs the design of M&E systems. Villanueva (2011) stresses that 'M&E for CCA is not about the *quantity* of indicators but about their *quality*'. The Livelihood Adaptation to Climate Change project in Bangladesh developed and applied a set of basic criteria and simple, measurable indicators to monitor and evaluate good practices for CBA in agriculture at the farm level. The main criteria include (1) agro-ecological suitability of good practices under location-specific conditions, (2) economic viability and social acceptability, (3) capacity to enhance climate resilience, (4) contribution to carbon footprint (Baas *et al.* 2010). A remaining challenge is, however, to motivate farmers and communities to engage over longer periods in M&E processes which induce additional work and do not trigger immediate benefits for the farmers themselves. M&E for CBA is more likely to succeed when linked to projects and action-research approaches, where outside assistance or incentives for M&E can be provided to communities or farmers.

Overall guidance of M&E related to adaptation in agriculture should be institutionalised within agricultural line agencies and led by a qualified M&E unit. At that level M&E results should be aggregated and used to inform regular agricultural development planning. The option of extension/research experts groups to guide M&E can be an alternative approach to institutionalise M&E.

Catalysing CBA through safety nets and better access to markets, financial services and insurance

Upscaling CBA in communities where the transition to resilient agriculture is challenged by extreme poverty and high levels of food insecurity requires incentives and complementary measures that meet immediate nutrition and income needs of poor and food-insecure farmers and their families. Governments can support CBA among the poorest farmers through innovative programmes that integrate social safety nets, livelihood-strengthening measures and risk transfer mechanisms (see Box 12.2). CBA processes can also be catalysed through innovative approaches that link small farmers to markets. For example, combining the WFP's purchasing power with the technical expertise of partners, WFP's Purchase for Progress (P4P) initiative offers smallholder low-income farmers opportunities to access agricultural markets and become competitive players in those markets. Through P4P, over 200 farmers' organisations have been contracted to sell food commodities to the WFP since 2008, and over 120,000 farmers, warehouse operators and small and medium traders have received support and training in organisation management, sustainable farming techniques, risk management, quality control and post-harvest handling (WFP 2012).

Safety nets that transfer food and cash to poor community members are not only essential in preventing the deterioration of food and nutrition security during climate-related disasters; food- and cash-for-work programmes or household income support measures can also allow poor farmers and community members to make investments in adaptive livelihood alternatives and promote the creation of community assets that are critical for CBA. For example, Ethiopia's Productive Safety Net Programme (PSNP) is a large-scale example of a safety net that prevents vulnerable households from selling livelihood assets in times of crisis and promotes CBDRR such as soil and water conservation on communal and household lands (World Bank 2011). To support the implementation of national food security policy and the PSNP, the government of Ethiopia partnered with the WFP and World Bank to improve its capacity to manage drought and flood risk by developing an integrated national risk management framework through the Livelihoods, Early Assessment and Protection (LEAP) project. LEAP links food security early warning to contingency planning, which is supported by a dedicated contingency finance mechanism. Using near real-time agro-meteorological data derived from satellites and automated weather stations, as well as livelihoods baseline and crop data collected through regular surveys, the LEAP early warning system monitors drought conditions, estimates needs, quantifies them, and triggers the funding needed to implement contingency plans to scale-up the PSNP. This allows for much earlier support to be provided to farmers, which in turn enables them to manage crises better and prevents them from taking actions that have a long term detrimental livelihood impact, such as distress migration or sale of productive assets (Hess and Balzer 2010).

Similarly, financial services, provided in partnership with the private sector, including tailor-made microfinance products or crop and livestock insurance, can facilitate the access of poor farmers to upfront investment costs of implementing adaptation measures and can thereby catalyse the adoption of adaptation practices. One innovative example of this type of service is the 'insurance-for-work' mechanism that allows poor, food-insecure households the opportunity to pay for insurance with their labour as part of a labour-based safety-net programme. With insurance, these households have better access to credit and reduced uncertainty from climate variability, which allows them to take risks and invest in adaptation strategies. This approach is now being scaled up by Oxfam America and the WFP through the R4 Rural Resilience Initiative, a five-year partnership that builds on the Horn of Africa Risk Transfer for Adaptation programme successfully implemented in Ethiopia's Tigray region. This programme broke new ground with its holistic approach in supporting cash-poor farmers and landless rural households to pay for their insurance with their own labour. R4 enables poor farmers to strengthen their food and income security by managing risks through a four-part approach: improving natural resource management (community risk reduction), accessing microcredit ('prudent' risk taking), gaining insurance coverage (risk transfer) and increasing savings (risk reserves). This programme has scaled up from 200 households in the first year to 13,000 households in 2011, and over the next five years will be scaled up to reach 500,000 people in Ethiopia, Senegal and two additional countries (WFP and Oxfam America 2011).

Box 12.2 CBA in the context of transitioning from chronic food insecurity to resilient livelihoods: the Karamoja Productive Assets Programme in Uganda

As part of Uganda's UN Joint Programme on Climate Change, the WFP is implementing the Karamoja Productive Assets Programme (KPAP), a large-scale food- and cash-for-work and asset-creation programme designed in consultation with the Office of the Prime Minister. Karamoja is part of the semi-arid, pastoralist belt of the Horn of Africa. It is the poorest and most marginalised region in Uganda. Droughts have become more frequent and more severe, in part as a result of climate change. Over time, frequent natural disasters, ongoing violence, severe environmental degradation and high poverty rates have eroded people's capacity to cope and left them dependent on food aid.

The launch of KPAP in 2010 reflected a shift towards addressing the underlying causes of recurrent crises. So far, 76,000 chronically food-insecure households (about 40 per cent of the population) participated in activities aimed at strengthening livelihoods and resilience to climate-related risks. These include both public works to create livestock watering points, land/soil conservation measures and reforestation, etc., supported by food or cash transfers, as well as household income support measures where community members receive training and inputs for drought-resistant crops, vegetable gardens, fruit orchards, gum Arabic cultivation, dairy production, energy-saving stoves, etc. All activities are chosen by the communities from an 'investment menu' of low technology and labour-intensive CBDRR and CBA activities that are suitable in the context of

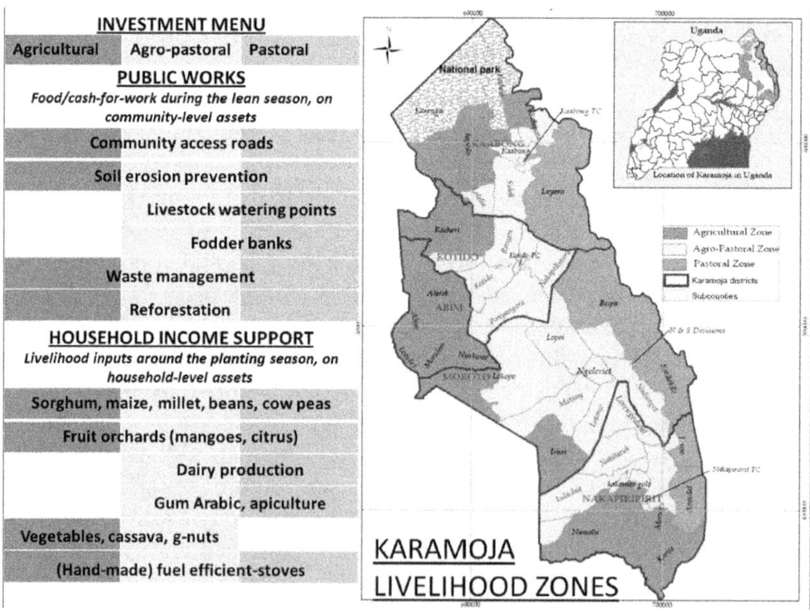

Figure 12.3 'Investment menu' of CBDRR and CBA activities in the context of the Karamoja Productive Assets Programme, Uganda.

each of the three different livelihood zones – agricultural, agro-pastoral and pastoral – in Karamoja.

KPAP is embedded in government institutions and policies. It is aligned with the government's Karamoja Action Plan for Food Security and an implementing arm of the government's Second Northern Uganda Social Action Fund. The programme also relies on strong technical and operational partnerships. FAO played a key role in developing the investment menu for KPAP and is advising district local governments, WFP and other implementing partners on a range of technical issues on the programme. This is essential for quality assurance, as well as harmonisation with FAO-supported agro-pastoralist field schools. Nine NGO partners also work closely with WFP to adjust geographical coverage and ensure adequate implementation capacity for scaling up.

Conclusions and way forward

Adaptation is a process incorporating initiatives that range from adopting and recommending local practices to putting the necessary institutional, policy mechanisms and technical capacities in place. CBA, at the interface between bottom-up adaptation initiatives and government-led interventions, is a suitable pathway to enhance capacities for adaptation at the local level. In most developing countries, agriculture, which is highly exposed to the impacts of climate change, is a major livelihood activity, particularly for the poorer segments of the population, and thus forms a central part of CBA. The agricultural sector has the potential to act as an 'engine' for its upscaling.

CBA in agriculture links short- and long-term perspectives. It builds on existing knowledge and practices, and is embedded in existing institutional structures. There is a wide range of existing tools and methods for DRR and development which, if adapted to the additional challenges caused by climate change, can also support CBA. Strong community networks are essential to ensure the information flow between farmers and agriculture support services. CBA in agriculture is most successful when farmers see tangible benefits and are willing and empowered to engage in sustained CBA processes.

Upscaling CBA in agriculture therefore requires supportive institutional and policy frameworks that catalyse its implementation. Agriculture sector agencies have a key role to play in sector-specific as well as cross-sectoral coordination, technical guidance and capacity building to enable CBA. Successful upscaling of CBA in agriculture depends on organisational and institutional learning and multi-stakeholder engagement at and across all levels. This needs to be complemented by planning mechanisms that ensure the strategic integration of CBA into regular agricultural and rural development planning.

Upscaling CBA in agriculture also requires incentives and services that complement and bring added value to existing community activities and structures. Of crucial importance is the strengthening of extension and research systems to guide and facilitate iterative and flexible planning and testing adjusted to agriculture planning cycles, including M&E to ensure effectiveness of climate-resilient and demand-driven practices. This process relies on systematic knowledge management, communication and awareness-raising methods, tools and products to enable continuous participation of communities and farmers in the design and implementation of CBA strategies and practices. In communities challenged by extreme poverty, the upscaling of CBA must be complemented through access to financial services, safety nets, social services and insurance that allow poor farmers to invest in alternative livelihood strategies.

Efforts in upscaling CBA in agriculture may, however, be hampered by the absence of political will and commitment to agriculture, as well as by poor coordination between agriculture and other sectors. Political instability, high turnover of governments, and very short-term funding horizons can hinder sustained support services for a long-term approach to reduce climate risk in agriculture. Upscaling efforts should therefore consider the governance context as well as the legal, social, economic and political environments that facilitate commitment, investments and provision of leadership in favour of CBA.

Notes

1 In the following sections the term *agriculture* is used according to FAO terminology, which includes also fisheries and forestry perspectives.
2 Drought-tolerant wheat varieties in Shandong Province, China.
3 Examples of CBA technologies and practices which addressed natural resource management include: reduced soil erosion through conservation tillage in Nepal (Ministry of Agriculture and Cooperatives Nepal 2010); increased water retention and reduced runoff and soil loss through combined system for rain water harvesting and soil conservation in Saint Lucia (Ministry of Agriculture, Food Production, Fisheries and Rural Development, Saint Lucia 2011); increased soil fertility and productivity by rapid composting of rice straw in irrigated lowlands in the Philippines (De La Torre 2010).

References

ADPC and FAO (2010) 'Adaptive learning in climate risk management: final report – reflections on capacity building of livelihoods adaptation to climate change (phase-ii) in Bangladesh' www.fao.org/climatechange/laccproject/en (accessed March/April 2012).

Agrawal, A., McSweeney, C. and Perrin, N. (2008) 'Local institutions and climate change adaptation'. In: *Social Development Notes: Community Driven Development; The Social Dimension of Climate Change*. World Bank, Washington, DC.

Ayers, J. and Anderson, S. (2011) Background paper prepared for the Experts Workshop on Participatory Monitoring and Evaluation for Community-based and Local Adaptation, 17–18 February, CARE UK/IIED.

Baas, S., Zanev, C., Hiepe, C. and Alacevich, T. (2010) 'Livelihood Adaptation Project to Climate Change, framework for analyzing, prioritizing, and recommending (or not) small technology options for replication in the context of adaptation to climate change', FAO, http://foris.fao.org/static/data/nrc/LACCII_evaluationframework_CCadaptationoptions.pdf (accessed March/April 2012).

BRRI-FAO (2009) 'Results and findings from BRRI-FAO collaboration under project *Improved Adaptive Capacity to Climate Change for Sustainable Livelihoods in the Agricultural Sector in Bangladesh*'. Final Synthesis Report, Dhaka.

CARE (2009) *Climate Vulnerability and Capacity Analysis Handbook*. www.careclimatechange.org/cvca/CARE_CVCAHandbook.pdf.

Centre for International Agricultural Development of China (CIAD) (2009) *Strengthening Disaster Preparedness of Agricultural Sector in China*, Beijing, China.

DAE Bangladesh and FAO (2008) 'Results and lessons from field testing of local adaptation options for agriculture in drought-prone areas of north-western Bangladesh 2005–2007'.

De La Torre, E. (2010) 'Detailed implementation guidelines, good practice options, lowland irrigated agro-eco zone', prepared for the Strengthening Capacities for Risk Management and Disaster Preparedness in Selected Provinces of the Philippines (Bicol Region). http://foris.fao.org/static/data/nrc/bicol/Report_Implementation_guidelines_lowlands.pdf (accessed March/April 2012).

Department of Agriculture, Cordillera Region of Philippines (2012) 'Compendium of good practices for climate change adaptation: experience from the Cordillera Region in the Philippines'.

FAO (2011a) 'Submission to the UNFCCC Secretariat on possible areas of further work on impacts, vulnerability and adaptation to climate change under the Nairobi work programme', http://unfccc.int/resource/docs/2011/smsn/igo/136.pdf (accessed March/April 2012).

FAO (2011b) *FAO Framework Program on Climate Change Adaptation*, FAO, Rome.

FAO (2010) *The State of Food Insecurity in the World*, FAO, Rome.

FAO (2009) *Report of the FAO expert meeting on How to Feed to the World in 2050*, FAO, Rome.

FAO (2008a) *Climate Change and Food Security: A Framework Document*, FAO, Rome.

FAO (2008b) *Community-Based Adaptation into Action: A Case Study from Bangladesh*, FAO, Rome.

FAO (2007) *The Global Information and Early Warning System on Food and Agriculture*, FAO, Rome.

Füssel, H.M. (2007) 'Vulnerability: a generally applicable conceptual framework for climate change research', *Global Environmental Change*, 17 (2), pp. 155–167.

Gero, A., Méheux, K. and Dominey-Howes, D. (2011) 'Integrating community based disaster risk reduction and climate change adaptation: examples from the Pacific', *Natural Hazards Earth System Sciences*, 11 (1), pp. 101–113.

GIZ (2011) 'Adaptation in agriculture', in *Adaptation to Climate Change with a Focus on Rural areas in India*, Deutsche Gesellschaft für Internationale Zusammenarbeit (GIZ) GmbH, New Delhi.

Government of Ecuador (2011) 'Enhancing resilience of communities to the adverse effects of climate change on food security, in Pichincha Province and the Jubones River Basin', Adaptation Fund Project by the Ministry of Environment in coordination with Ministry of Agriculture, Livestock, Aquaculture and Fisheries, Commonwealth of the River Jubones Basin and Provincial Government of Pichincha, implemented by WFP.

Harris, K., Seballos, F., Silva Villanueva, P. and Curmi, P. (2012) *Changing Climate, Changing Disasters: Pathways Towards Integration*, Brighton, IDS.

Hess, U. and Balzer, N. (2010) 'Climate change and weather-risk management: evidence from index-based insurance schemes in China and Ethiopia', in: Omamo, S.W.,

Gentilini, U. and Sandstrom, S., *Revolution: From Food Aid to Food Assistance. Innovations in Overcoming Hunger*, WFP, Rome.

Hooper, M., Jafry, R., Marolla, M. and Phan, J. (2002) 'Scaling-up community efforts to reach the MDGs: an assessment of experience from the Equator Prize', in D. Roe (ed.), *The Millennium Development Goals and Conservation: Managing Nature's Wealth for Society's Health*, IIED, London, www.iied.org/Gov/mdgs/publications.html (accessed March/April 2012).

Huq, S. and Reid, H. (2007) 'A vital approach to the threat climate change poses to the poor', in *Community-based Adaptation to Climate Change: An IIED Briefing*, IIED, London.

IFAD (2010) *Rural Poverty Report 2011*, IFAD, Rome.

IIED (2009) 'Participatory learning and action 60', in *Special Issue on Community-Based Adaptation to Climate Change*, IIED, London.

IPC Global Partners (2008) *Integrated Food Security Phase Classification Technical Manual*, Version 1.1, FAO, Rome.

IPCC (2012) 'Managing the risks of extreme events and disasters to advance climate change adaptation (SREX), summary for policymakers', in *A Special Report of Working Groups I and II of the Intergovernmental Panel on Climate Change*, Cambridge University Press, Cambridge.

Jules, C. (2011) 'Communications strategy report: enhancing capacities for disaster risk mitigation in agriculture, fisheries & forestry, Saint Lucia'. Project Report.

Macchi, M., Gurung, A.M., Hoermann, B. and Choudhary, D. (2011) *Climate Variability and Change in the Himalayas, Community perceptions and responses*, International Centre for Integrated Mountain Development, Kathmandu.

Ministère de l'Agriculture, des Ressources Naturelles et du Développement Rural (MARNDR), République d'Haïti et FAO (2007) 'Bonnes Pratiques pour la Gestion des Risques et des Désastres dans l'Agriculture' dans le projet *Assistance à l'Amélioration de la Préparation Locale face aux Urgences et Catastrophes de Nature Agricole dans les Pays de la Caraïbe Hautement Exposés aux Dégâts Associés aux Cyclones*. ftp://ftp.fao.org/docrep/fao/011/ak178f/ak178f00.pdf (accessed March/April 2012).

Ministry of Agriculture and Cooperatives, Nepal, FAO (2010) 'Plan of action for climate change adaptation and disaster preparedness in agriculture 2011–2015'.

Ministry of Agriculture and Fisheries, Belize (2011) 'Plan of action for disaster risk reduction in agriculture'. www.fao.org/climatechange/29975-0236c2c8f876fc3e77a354a7a4585df01.pdf (accessed March/April 2012).

Ministry of Agriculture and Rural Development Ethiopia (2005) *Community Based Participatory Watershed Development: A Guideline*, Ministry of Agriculture and Rural Development Ethiopia, Addis Ababa.

Ministry of Agriculture, Food Production, Fisheries and Rural Development, Saint Lucia (2011) 'Enhanced capacities for disaster risk mitigation in agriculture, fisheries and forestry', Project Report.

Ministry of Agriculture, Grenada (2008) 'Good agricultural practices for climate risk management in Grenada', in *Assistance to Improve Local Agricultural Emergency Preparedness in Caribbean Countries Highly Prone to Hurricane Related Disasters* www.fao.org/fileadmin/templates/tc/tce/pdf/Grenada_draft_final_report_May_2008.pdf (accessed March/April 2012).

Mitchell, T. and van Aalst, M.K. (2008) 'Convergence of disaster risk reduction and climate change adaptation', DFID.

Mitchell, T., Ibrahim, M., Harris, K., Hedger, M., Polack, E., Ahmed, A., Hall, N.,

Hawrylyshyn, K., Nightingale, K., Onyango, M., Adow, M. and Sajjad Mohammed, S. (2010) 'Climate smart disaster risk management', in *Strengthening Climate Resilience*, IDS, Brighton.

Nedessa, B. and Wickrema, S. (2010) 'Disaster risk reduction: experience from the MERET project in Ethiopia', in: Omamo, S.W., Gentilini, U. and Sandstrom, S., *Revolution: From Food Aid to Food Assistance. Innovations in Overcoming Hunger*, WFP, Rome.

Parry, M., Evans, A., Rosegrant, M. and Wheeler, T. (2009) *Climate Change and Hunger: Responding to the Challenge*, WFP, IFPRI, New York University Center on International Cooperation; Grantham Institute at Imperial College London; and Walker Institute at University of Reading. www.ifpri.org/sites/default/files/publications/wfp_fightingcchunger.pdf (accessed March/April 2012).

Schipper, L. and Pelling, M. (2006) 'Disaster risk, climate change and international development: scope for, and challenges to, integration' *Disasters*, 30 (1), pp. 19–38.

University of the Philippines Los Baños (2011) 'Local knowledge and tools for assessing vulnerabilty of the agricultural systems to changing climate: the case of Ifugao and Benguet Provinces (synthesis report)', in *Strengthening Philippine Institutional Capacity to Adapt to Climate Change, 2nd Mid-Term Progress Report*, UPLB Foundation Inc., Laguna.

Urquhart, P. (2010) *WFP and Climate Change: A Review of On-going Experience and Recommendations for Action*, World Food Programme, Rome.

Venton, P. and La Trobe, S. (2008) 'Linking climate change adaptation and disaster risk reduction', Tearfund and Institute of Development Studies.

Villanueva, P.S. (2011) 'Learning to ADAPT: monitoring and evaluation approaches in climate change adaptation and disaster risk reduction – challenges, gaps and ways forward', in *Strengthening Climate Resilience Discussion Paper 9*, IDS, Brighton.

Uvin, P. and Miller, D. (1996) 'Paths to scaling-up: alternative strategies for local non-governmental organizations', *Human Organization*, 55 (3), pp. 344–354.

WFP (2012) *P4P Purchase for Progress: A Primer*, World Food Programme, Rome.

WFP (2011a) *Change and Hunger: Towards A WFP Policy on Climate Change*, World Food Programme, Rome.

WFP (2011b) *WFP Policy on Disaster Risk Reduction and Management: Building Food Security and Resilience*, World Food Programme, Rome.

WFP and Oxfam America (2011) 'R4 Rural Resilience Initiative: partnership for resilient livelihoods in a changing climate', http://documents.wfp.org/stellent/groups/public/documents/communications/wfp240451.pdf (accessed March/April 2012).

World Bank (2011) 'Social protection and climate resilience', report from an international workshop held in Addis Ababa, 14–17 March 2011, organized by the World Bank in collaboration with IDS, DFID and UNECA. http://siteresources.worldbank.org/EXT-SOCIALDEVELOPMENT/Resources/244362l232059926563/5747581–1239131985528/WBSocProtec_Final.pdf (accessed March/April 2012).

World Bank (2008) 'Agriculture for development', in *World Development Report 2008*, World Bank, Washington, DC.

13 Community participation in urban adaptation to climate change

Potential and limits for community-based adaptation approaches

Katarina Soltesova, Anna Brown, Ashvin Dayal and David Dodman

Rapid urbanisation is often associated with both prosperity and new challenges in improving the well-being of low-income urban residents. Although cities are often conceived of as engines of economic growth and cultural development, more than one billion urban residents worldwide live in conditions of inadequate shelter, poor access to basic services, insecurity of land tenure and uncertain livelihoods. Climate change will bring gradual change and extreme events that compound the existing risks facing these people. Dominant responses to climate change in urban areas have tended to focus on the role of municipal or national governments, sometimes in association with private sector partners (Bulkeley and Betsill 2005; Anguelovski and Carmin 2011; Corfee-Morlot *et al.* 2011). However, in recent years a growing number of development NGOs and international organisations have started to engage with efforts to assist towns and cities to adapt to climate change, with increasing interest in the potential of community-based adaptation (CBA) to build adaptive capacity among low-income urban residents. These approaches draw on aspects of governance, such as local decision-making and ownership, and utilise a range of tools and approaches intended to integrate community participation into urban planning. However, there has been relatively little discussion about the ways in which the particular nature of urban societies and economies might shape the implementation of CBA, or how to scale-up lessons learned within and across cities.

In urban areas in low- and middle-income nations, CBA develops amidst pressures and opportunities brought about by processes inherent to rapid urbanisation. The scale and rate of urban population growth often increases urban risk in ways which require policy support and collective action across sectors and institutional scales. Urbanisation of hazard-prone areas and disregard of existing adverse environmental conditions contributes to vulnerability in many new neighbourhoods, while rapidly developing residential or industrial areas can undermine the sustainability of existing neighbourhoods by altering ecosystems and placing excessive stress on infrastructure. Increasing densities within urban settlements of all sizes and urban sprawl alike add further strain on local infrastructure and accelerate environmental degradation, highlighting the need for city governments to strengthen provision of basic services, address pollution of all nature and promote the preservation of urban ecosystems, as well as incorporate mitigation measures in their future planning. (Romero-Lankao and Dodman 2011). For example, Dhaka's traditional urban surface water bodies, which formerly served to regulate water flows within and beyond the residential and commercial areas, are now diminishing in size and deteriorating in quality due to illegal filling and chemical/sewage contamination. This process decreases both the capacity of the city to counter potential major floods and puts at great risk poor households who rely on surface and well water and live in the immediate vicinity of the remaining lakes.

Urban CBA is well suited to engage with urban political ecology to identify entry points likely to alleviate imbalances of urban systems. In responding to uneven urban development, it has the potential to act as a bridge between local, often informal, arrangements and formal institutions structuring urban development. Enquiring into the nature of urban vulnerability highlights not only the biophysical condition of cities, but also draws attention to the socio-political and historical context of urban development as pertinent to specific local communities.

Complex urban relationships, including organisational networks, facilitate contact between individuals and groups, support sharing of information and contribute to the development of collective capacity to mobilise in order to benefit more fully from the diversity of opportunities presented within the city. Because of this, collective solutions will underpin effective adaptation to climate change in urban areas. The ways in which local communities organise and mobilise are central to their capacity to negotiate priorities and pursue strategies striving to increase their political leverage and commitment from other urban actors, while long-term multi-stakeholder partnerships aimed at opening avenues for low-income urban residents to engage in decision-making processes both at the neighbourhood and city levels are key to sustainable and environmentally just urban development.

This chapter uses examples of community responses to climate variability and change from India, Vietnam and Bangladesh to analyse the potential for CBA to enable meaningful participation in responding to climate change risks in urban areas. It examines the key characteristics of urban resilience and adaptation, and incorporates a strong focus on collective action and institutionalisation of community-led risk reduction. Experiences from a range of stakeholders in Asian cities demonstrate the potential for collective engagement of low-income urban residents and communities in responding to threats posed by climate change. As illustrated through the case studies, collective mobilisation relies on long-term commitment and strategic action, reaching outward to engage actors from among local municipal governments, policy-makers, other civil society groups and the private sector.

The first section of the chapter provides a brief literature review that frames the key issues that are considered during the examination of the case studies.

Three case studies illustrate ways in which urban residents within and across cities organise and act through partnerships and networks of organisations, engaging with processes of scaling up community involvement at city-wide, national and regional levels. The chapter closes with a discussion of pathways to conceptualise urban resilience, urban systems and upscaling of CBA.

The potential of CBA for increasing urban resilience

Urban vulnerability encompasses a range of dimensions which condition the impact of climate change on urban societies. The Intergovernmental Panel on Climate Change (IPCC) defines vulnerability as determined by the exposure of a system, its sensitivity and its adaptive capacity (Parry *et al.* 2007). An emphasis on adaptive capacity highlights how urban actors and institutions respond to climate change (Romero-Lankao and Qin 2011; Schensul and Dodman 2010) and provides a key entry point at which community-level action can reduce vulnerability. Vulnerability of individuals and groups varies given their knowledge, resources, assets or political leverage. A capacity to muster social capital, including institutional and political capital (McLeod 2002), or to adapt one's understanding and behaviour in response to new threats is central to the understanding of vulnerability as a negotiated and dynamic process, which takes place within a particular neighbourhood, but also at city level where it requires sustained dialogue between communities, local governments and municipal officers.

Several key principles for promoting sustainable and socially just urban development are equally relevant for reducing vulnerability to climate variability and change in urban areas, particularly around community mobilisation and the opening of spaces for participation in urban decision-making processes. Many elements of resilience are consistent with the provision of basic urban services and the effective delivery of key urban functions – around issues including provision of water and sanitation, solid waste management, public health and disaster risk reduction – by municipal authorities (Satterthwaite *et al.* 2007). However, while recognising that ensuring urban physical infrastructure (including roads, drainage networks and communications infrastructure) is able to withstand the impacts of physical hazards is a key component of resilience, other elements stand out when considering cities' ecological sustainability; urban disaster management and disaster risk reduction, urban and regional economics, and governance and institutions are also critically important (Leitchenko 2011).

Examples of local mobilisation around issues including community savings, enumerations of households undertaken by communities themselves (Karanja 2010) and community land acquisition highlight the centrality of land and shelter as key elements of resilience for low-income communities (Dodman *et al.* 2010; Jabeen *et al.* 2010). Partnerships developed over time between organised communities and other urban actors provide avenues to compare experiences, share knowledge, muster political support and influence policy so as to tap into funding schemes, change the organisational structures of relevant agencies in view of accommodating community needs and influence implementation of projects (Levy 1996). Partnerships across scales of urban governance provide opportunities for low-income urban residents to engage in debates and policy-making on issues extending beyond immediate community boundaries, such as land use or degradation of urban ecosystems.

A political ecology approach provides a strong theoretical basis for understanding the complex processes that are constantly being negotiated within the context of the urban environment. This analysis is based on three core premises: environmental change is not independent of human influence; unequal socio-economic relations are mirrored through environmental 'goods' and 'bads' at multiple spatial and temporal scales; and environmental change threatens the status quo of dominant groups and will thus require that different actors engage in resolving environmental problems through synergies at the local, regional and global scale (Bryant and Bailey 1997). Addressing environmental problems requires engaging with their ultimate rather than proximate causes. Similarly, approaches for adapting cities to climate change can also stress a transformative agenda that re-shapes the cultural models influencing societies, economies and their institutions (Pelling 2011). If CBA is to make meaningful changes in the prospects of urban households and communities to respond to climate change, it needs to engage with a transformative agenda that seeks not only to reduce the surface causes of vulnerability, but also to address underlying social and political drivers (Dodman and Mitlin 2011).

Case studies: community responses to climate change in India, Vietnam and Bangladesh

The above concerns and prospects can all be observed in contexts where urban CBA – although admittedly embryonic in nature – is already taking place. The three case studies that follow examine the value of CBA in responding to the needs, experiences and ambitions of urban residents, including the urban poor, who are often most severely affected by the extreme and slow onset impacts of climate change.

Case study 1: aspects of vulnerability in Gorakhpur

The case of Gorakhpur illustrates the use of participatory planning tools and a systems analysis approach to understand local vulnerability and key climate impacts. It generates a collective understanding of actors and entry points for action through partners, some of which come together to address specific thematic areas, while others comprise a broader pool of experience, knowledge and expertise and political leverage.

Gorakhpur is located in the mid-Gangetic plains in India, and counts among the region's fastest growing cities. The development of basic urban services lags behind the city's growth, as does conservation of natural flood storage and drainage bodies, waste disposal infrastructure, drainage and sewer infrastructure, which is provided to only 22 per cent of the city (ACCCRN 2009). Given the city's varied topography and dysfunctional urban infrastructure, it is anticipated that changes in the variability and intensity of rainfall will exacerbate risks associated with flooding and waterlogging, which have already risen to be the most acute of the city's problems, affecting the deteriorating water quality and spread of water- and vector-borne diseases. The city is subject to chronic waterlogging for 3–4 months each year (ACCCRN 2009; Wajih *et al.* 2010). The 33 per cent of the city's population living in informal slum settlements (Wajih *et al.* 2010) are among the most vulnerable groups in the city.

Starting in 2008, initiatives undertaken by the Asian Cities Climate Change Resilience Network (ACCCRN) and led by the Gorakhpur Environmental Action Group (GEAG) have engaged in efforts to understand vulnerability across the city and to examine possible measures to increase resilience. Through direct engagement with communities, local government and representatives from local academic institutions, the process led to the development of a city resilience strategy. Building local resilience is understood as requiring interventions that support knowledge about current conditions, a shared understanding of possible future impacts, demonstrated precedents and development of the capacity of organisations to drive behavioural, political and institutional change. To date, the resilience planning process took place through three shared learning dialogue (SLD) sessions, which helped integrate findings from a vulnerability assessment, climate scenarios and sector studies, as well as an understanding of projected urban growth to inform the City Resilience Strategy. Leading the process of participatory vulnerability assessment, GEAG employed a systems analysis approach and participatory tools to identify 'causal loops' (e.g. diagrams visualising causes and sub-causes of waterlogging) which allowed input from multiple stakeholders as well as identification of the relationship between facts, problems, distinct actors and possible entry points leading to solutions at different scales of the urban system. Participatory vulnerability assessment and priority setting constituted the micro-resilience planning process. A shared understanding that a combination of physical, infrastructure, behavioural and institutional (governance) requirements undermine the ability of complex urban systems to meet the needs of the city's poor communities guided the stakeholders in identifying possible responses as well as policy and implementation partners.

Implemented projects include community-level drainage maintenance, which stems from ward-level micro-resilience plans. GEAG also initiated a pilot community waste collection intervention aimed at preventing drain blockage and changing the informal practice of disposing of waste in city lakes. This raised awareness about eutrophication processes and the benefits of the lakes' flood retention capacity. In the process of developing the City Resilience Strategy, linkages have been highlighted between health and the inadequate attention paid to drainage, solid waste disposal and waterlogging. The City Resilience Strategy also identifies community surveillance as one entry point to help anticipate and better prepare for health outbreaks, which may serve as an intervention that stakeholders pursue in the future to address the city's health systems directly.

The urban self-governance provision made possible through the 74th Amendment of the constitution has not yet fully extended to Gorakhpur. Entry points for direct government action in the city are thus weaker and require solutions rooted in strong local partnerships. GEAG and the other local institutions therefore seek to drive resilience-building measures through mobilisation of and engagement with civil society and community groups. The development of a City Resilience Strategy, which includes possible resilience-building measures that focus on different scales, populations and sectors of the city, has further enabled municipal authorities and civil society actors to approach donors and establish new links with government ministries, opening channels for vertical scaling up of adaptation and risk reduction processes across urban scales.

Case study 2: partnerships for resilience in urban Vietnam

Developing city climate change resilience strategies in three Vietnamese cities – Da Nang, Can Tho and Quy Nhon – exemplify a process of bridging core municipal priorities to strengthen resilience, accounting for the everyday realities of communities facing distinct urban conditions and associated risks, with national and regional government policy and programming priorities.

Due to its low elevation and coastal position, Vietnam is projected to be one of the top five countries most affected by climate change (ACCCRN 2009). Can Tho City, the largest city in the Mekong River Delta, faces adverse climate change impacts such as increased temperatures, changing patterns and intensity of monsoon rainfall, drought and sea-level rise, which will negatively impact aquaculture and rice-growing areas. Da Nang City is located on the central coast of Vietnam and is prone to typhoons and rain-induced flooding, both of which result in flash floods, landslides and soil erosion. Prolonged drought will jeopardise agriculture production and the city's water supply. Risk of saline water intrusion ranks equally high. Finally, the city of Quy Nhon comprises low-lying coastal land, subject to frequent flash floods and river flooding, particularly during the monsoon season, and exacerbated by runoff from mountains running alongside the city's western edge.

The National Target Program to Respond to Climate Change presents a key opportunity for adaptation upscaling in urban Vietnam, calling for each municipality to generate a climate action plan. Distinct urban sectors will face disparate climate impacts, though no one sector can singlehandedly bring about resilience for a city. National policy discussions, convened by the National Institute for Science and Technology Policy and Strategy Studies (NISTPASS) and the Institute for Social and Environmental Transition (ISET), thus gathered representatives from a cross-section of ministries – including Planning and Investment, Environment and Natural Resources, and Construction – to raise the need for municipalities and urban communities to plan in the face of uncertainty. Success of the National Target Program, though still nascent in its guidance and implementation, will require cross-sector and collaborative partnerships across the national, city and community levels. An experiment that could prove instrumental in this effort is the creation of intra-governmental coordination offices – Climate Change Coordination Offices (CCCO) – which have been established in each of the three ACCCRN cities.

City resilience strategies have been developed for the three cities using an iterative process grounded in participatory methodology and system-thinking championed by institutions engaged in ACCCRN. The process has been developed and facilitated by ISET, a non-profit organisation with strong technical capacities in climate science, process facilitation and research. At the national level, NISTPASS, a leading national policy institute, has provided critical leadership and guidance to foster the cooperation of national ministries and their respective technical institutions in undertaking climate change projections and analyses. Challenge to Change (CtC) is an important NGO partner – particularly in the Vietnam context where government plays a dominant role, leaving often limited space for civil society groups and community-based organisations (CBOs). CtC undertook community-based hazard and vulnerability assessments and a pilot project in community/government co-management of local mangrove forests.

In all three cities, a multiplicity of stakeholders who would otherwise operate within narrowly defined professional and organisational boundaries, exchanged views and knowledge by means of SLDs.[1] Expert and community meetings led to the development of vulnerability assessments, sector studies and pilot projects, which provided space for the inclusion of communities traditionally excluded from planning and political processes. Distinct groups have been identified as particularly vulnerable during early meetings organised by ISET, CtC, municipal partners, community groups and local technical experts. The most vulnerable groups in the three cities include recent migrants who lack residential registration and are not entitled to compensation; communities formally resettled from coastal and flood plain sites; women, children and other childcare providers; and low-income urban residents, many of whom rely on the fisheries sector.

In a period of two years, participatory vulnerability assessments were followed by efforts to raise awareness among youth and the private sector, training sessions in community-based disaster risk management, detailed sector studies and, finally, the drafting of city resilience strategies highlighting key priority areas and projects that each city might take to strengthen resilience to climate change impacts. Crucial throughout the process has been a focus on the development of trust among government officials, professional groups, NGOs and community representatives. Trust is understood as a key component of social capital, allowing for social learning, organisational development and community mobilisation.

Case study 3: networking communities in Korail, Dhaka

Community organisation in Dhaka's Korail settlement illustrates ways in which urban adaptation to climate change takes place without a formulated adaptation strategy, but it develops underpinned by long-term sustained negotiation of socio-political relations. Low-income urban communities actively create and tap into networks of actors operating at all scales of urban governance. Climate and environmental hazards compounded by a lack of tenure security, inadequate housing, a lack of basic services and risk-prone changes in land use across Dhaka require concerted action by communities living in informal settlements and their partner organisations.

Dhaka's largest informal settlement counts a population of over 100,000 (CUS 2006). While adjacent to Dhaka's most prominent business and diplomatic zones, Korail is surrounded by the severely polluted Gulshan Lake and remains spatially disconnected from other neighbourhoods. Given its prime location, ideas for its redevelopment have been debated for decades and inspire frequent eviction threats. Korail's land belongs to four government agencies, while informal private landlords manage most of the existing housing units (CUS 2006). Rent seeking results in extreme density of the built environment, limited provision and maintenance of basic infrastructure and scarce open space. Key concerns identified by Korail's residents include frequent waterlogging, extreme indoor and outdoor temperatures and open-fire hazards. All the above pose high risks to human health and safety, impacting predominantly women, children and the elderly, but often implying a loss of key assets, inability to attend work and high costs for medical treatment incurred by the households.

Adaptation in Korail takes place at the household and community levels. Individual households have houses insulated with cardboard, raised beds and furniture; some have houses reinforced to withstand heavy monsoon rains and altered structures to allow for air circulation and better drainage. Most changes require substantial investment and long-term negotiation with informal landlords, none of whom are accountable for the quality of housing and services provided in the settlement. Community responses to climatic hazards in Korail are developed through a dense network of relationships and partnerships nested within grassroots CBOs, NGOs, municipal bodies, think-tanks and universities. Partners contribute technical expertise, support local CBOs in mobilising, exerting political pressure and advocating for policy and procedural change. In Korail, the existence of these informal networks spans two decades and responds to key urban development pressures and causes of vulnerability, namely the need for provision of basic services and tenure security.

The Organisation of the Urban Poor in Dhaka (NDBUS) is a federation of CBOs created in 1999, working for the welfare of low-income communities. In its early years, NDBUS community leaders, including those in Korail, drew on the work of a coalition of development NGOs (the Coalition for the Urban Poor, CUP) which advocates for housing quality, poverty reduction and sustainable urban planning. Partnering with an urban research think-tank, the Centre for Urban Studies, NDBUS leaders equally engage in policy advocacy, making a case for environmentally just and pro-poor urban development. Such initiatives complement acquisition of political (voting) rights for urban poor populations and demands made for greater protection against forced evictions. NDBUS members currently utilise institutional capital acquired over time through work with partners to negotiate options for long-term resettlement plans.

Despite the current lack of secure tenure, Korail CBOs, supported by the Vulnerable Poor Health Centre (DSK), partner in construction and maintenance of drainage and basic sanitation in Korail. Moreover, they shared the success of having negotiated provision of drinking water by the state water authority (WASA). Such precedents are key to further provision of drinking water, the lack of which will be exacerbated by climatic (seasonal drought), environmental (contamination of well water through both dysfunctional sewage/drainage infrastructure and worsening pollution of the surrounding surface water bodies) and socio-economic (over-extraction of groundwater by middle-class residents and the private sector) processes.

Collective learning within and across networks has resulted in increased trust among partners, change in the perception of risk and CBOs' capacity to manage environmental risk. CBOs' demonstrated financial and service management capacity has led municipal authorities to emulate DSK's community management model. Existing CBOs have been instrumental in rolling out the implementation of a major national urban poverty reduction programme (UPPR),[2] which seeks to foster participatory planning. Use of participatory tools and pilots such as channelling of urban development funds directly to communities (acting as managers and procurers) increased the CBO leaders' political capital and capacity to juggle organisational identities strategically, so as to maximise leverage and demand inclusion in municipal governance structures. Future opportunities are presented through the national Comprehensive Disaster Management Programme, which aims to mainstream disaster risk reduction across national and municipal bodies complemented by community participation. A future challenge remains bridging of CBOs and the multiple professional and policy-making bodies who influence Dhaka's environmental planning and development policy at the city level. It is here that periodic collaboration and partnerships sustained over the past two decades among CBOs, NGOs, specific service providers, academic institutions and municipal officers facilitate collaboration between CBOs and urban specialists whose remits and professional traditions would otherwise provide few obvious avenues for community-based planning.

Urban CBA: key features and prospects for upscaling

These examples from India, Vietnam and Bangladesh demonstrate several key features of community organisation in low-income urban settlements that are relevant for understanding the potential of CBA to build resilience to climate change in urban areas.

First, urban systems are intricately linked with and interdependent on each other. Hard infrastructure supporting urban sectors such as water provision, drainage, sewage or transportation rely not only on technological solutions but also on soft management skills such as the ability of urban stakeholders to negotiate adequate provision, access to and costs of these facilities; and the functioning of ecosystem services within and outside the city. In Dhaka, poor communities not only suffer from inadequate living conditions caused by lack of basic services, tenure insecurity and inadequate shelter, severely impacting on the residents' health, but their livelihoods greatly depend on the city's inadequate transportation infrastructure – rickshaw drivers find their work increasingly more exhausting in the boiling heat of the city and increasingly waterlogged streets. The physical environment – including canals and lakes – has a regulatory and protective role, but can also function to provide transport routes and crucial ecosystem services (livelihoods, leisure, natural resources) that are particularly valuable for poor households. The concept of a rural–urban continuum also contributes to the understanding of urban systems. Cities are perpetually subject to dynamic pressures created by climatic and environmental processes occurring within their rural hinterlands. Dependence on external natural resources is but one important aspect of the relationship. People's patterns of migration and mobility are often seen as a possible indicator of system imbalances, but are more productively understood as key responses to environmental and non-environmental transformations and pressures (Tacoli 2009). As such, they can offer opportunities for the development of important coping strategies (during specific seasons, children are often sent to relatives living in safer rural areas), or facilitate knowledge for household resilience (design of shelter in some of Dhaka's informal settlements copies regional models adapted to withstand strong wind and rain).

Second, low-income households and communities in urban areas have already demonstrated their capacity to cope and innovate in response to climatic shocks and stresses. Rather than replacing these capacities with new processes and actions, urban CBA has the potential to strengthen them and to act upon factors which determine the extent to which communities can act autonomously or can reach out to seek and demand support from actors who constitute a broad institutional landscape. As seen in Korail, dwellings have been modified and reinforced to improve circulation of air and to withstand heavy rains at a household level, while pavements with drainage have been agreed as a community priority and have been constructed by community members with funds from a national programme. These are not isolated responses, but take place within a context where both formal and informal networks have developed over years and decades which drive collective responses to a range of urban development pressures. Although these platforms are site-specific, acknowledging them has the potential to provide existing and new actors with a solid foundation for the development and implementation of pro-poor climate change adaptation strategies.

Third, approaches such as the Shared Learning Dialogues used by cities in the Asian Cities Climate Change Resilience Network can be used to ensure meaningful engagement by traditionally marginalised groups in planning and decision-making processes to strengthen resilience to climate change. In Gorakhpur, these have enabled a wide range of stakeholders to identify the causes and sub-causes of waterlogging and appropriate responses to this, including community maintenance of drainage; in Vietnam these have provided space to include the perspectives of recent migrants, resettled communities, women and children and urban poor groups in developing vulnerability assessments, pilot projects and future planning. These approaches offer the potential to link CBA efforts to city-wide analysis and planning processes that are deliberately focused on building urban climate change resilience, and that provide at the same time an opportunity to deepen social and political processes while engaging the scientific, planning and governance communities.

Taken together, these lessons are relevant for considering the ways in which CBA can be upscaled in towns and cities. Partnerships of communities organised in CBOs with NGOs, technical experts and government officials are instrumental in developing a shared understanding of causes of local vulnerability and risk associated with climate change, but also provide a space for communities and community leaders to engage strategically with other actors, formally or informally, as they plan ahead. Likewise, they are crucial in the creation of entry points leading to realistic long- and short-term solutions. Horizontal federations of CBOs have proven highly successful in enabling marginalised communities to tap into the social and political capital of others who share similar concerns. The examples from India and Vietnam illustrate how entry points which comprise co-operation across sectoral agencies can lead to responses targeting the systemic nature of urban climate and environmental risk. Civil society organisations further play a pivotal role in strengthening public awareness and advocacy, engaging in debates on urban well-being, sustainability and environmentally just urban development. Scaling up CBA in urban areas therefore encompasses a multitude of possibilities for horizontal and vertical partnerships, which aim to create trust and long-term collaboration. In the short term, the process requires genuine commitment to bring planning and implementation of projects and urban interventions down to the community level so as to reflect the needs, experiences and ambitions of the urban poor. Careful tuning of organisational structures, procedures and methods will also be vital for CBA to fulfil its potential. Finally, extending beyond localised project delivery, stakeholders engaging in the process of CBA should engage more fully with urban networks in order to inform and inspire incremental (yet holistic) policy change, and emphasise the need for more socially and environmentally just urban centres.

Note

1 Shared Learning Dialogues (SLD)
2 Urban Partnership for Poverty Reduction 2007–2015 (UPPR) initiative

References

ACCCRN (Asian Cities Climate Change Resilience Network) (2009). *Responding to the Urban Climate Challenge*. ISET, Boulder, CO.

Anguelovski, I. and Carmin, J. (2011). 'Something borrowed, everything new: innovation and institutionalization in urban climate governance', *Current Opinion in Environmental Sustainability* 3: 169–175.

Bryant, R. and Bailey, S. (1997). *Third World Political Ecology*. London, Routledge.

Bulkeley, H. and Betsill, M. (2005). 'Rethinking sustainable cities: multilevel governance and the "urban" politics of climate change', *Environmental Politics* 14(1): 42–63.

Corfee-Morlot, J., Cochran, I., Hallegatte, S. and Teasdale, P. (2011). 'Multilevel risk governance and urban adaptation policy', *Climatic Change* 104: 169–197.

CUS (Centre for Urban Studies) (2006). *Slums of Urban Bangladesh: Mapping and Census, 2005*. Dhaka, Centre for Urban Studies.

Dodman, D. and Mitlin, D. (2011). 'Challenges for community-based adaptation: discovering the potential for transformation', *Journal of International Development*, DOI: 10.1002/jid.1772.

Dodman, D., Mitlin, D. and Rayos Co, J. (2010). 'Victims to victors, disasters to opportunities: community driven responses to climate change in the Philippines', *International Development Planning Review* 32(1): 1–26.

Jabeen, H., Allen, A. and Johnson, C. (2010). *Built-in resilience: Learning from Grassroots Coping Strategies to Climate Variability. Proceedings of the 5th Urban Research Symposium*. World Bank, Marseilles.

Karanja, I. (2010). 'An enumeration and mapping of informal settlements in Kisumu, Kenya, implemented by their inhabitants', *Environment and Urbanization* 22(1): 217–239.

Leichenko, R. (2011). 'Climate change and urban resilience', *Current Opinion in Environmental Sustainability* 3(3): 164–168.

Levy, C. (1996). 'The process of Institutionalising gender in policy and planning: the web of institutionalisation', *DPU Working Paper Series*, Working Paper No. 74.

McLeod, R. (2002). 'Research on bridging the finance gap in housing and infrastructure and the development of CLIFF'. Homeless International.

Parry, M., Canziani, O., Palutikof, J., van der Linden, P. and Hanson, C. (eds) (2007). *Climate Change 2007: Impacts, Adaptation and Vulnerability. Working Group II Contribution to the 4th Assessment Report of the Intergovernmental Panel on Climate Change*. Cambridge, Cambridge University Press.

Pelling, M. (2011). *Adaptation to Climate Change: From Resilience to Transformation*. London, Routledge.

Romero-Lankao, P. and Dodman, D. (2011). 'Cities in transition: transforming urban centers from hotbeds of GHG emissions and vulnerability to seedbeds of sustainability and resilience', *Current Opinion in Environmental Sustainability* 3(3): 113–120.

Romero-Lankao, P. and Qin, H. (2011). 'Conceptualizing urban vulnerability to global climate and environmental change', *Current Opinion in Environmental Sustainability* 3(3): 142–149.

Satterthwaite, D., Huq, S., Reid, H., Pelling, M. and Romero Lankao, P. (2007). 'Adapting to climate change in urban areas: the possibilities and constraints in low- and middle-income nations'. Humana Settlements Discussion Paper – Climate Change and Cities 1. International Institute for Environment and Development.

Schensul, D. and Dodman, D. (2010). 'Populating adaptation: incorporating demographic dynamics in climate change adaptation policy and practice'. Unpublished paper, presented at UNFPA/IIED workshop 'Population Dynamics and Climate Change II: Building for Adaptation', Mexico City, October.

Tacoli, C. (2009). 'Crisis or adaptation? Migration and climate change in a context of high mobility', *Environment and Urbanization* 21(2): 513–525.

Wajih, S., Singh, B., Bartarya, E. and Basu, S. (2010). *Towards a Resilient Gorakhpur*. Gorakhpur, Asian Cities Climate Change Resilience Network/Gorakhpur Environmental Action Group.

14 Gender in scaling up community-based adaptation to climate change

Helena Wright and Achala Chandani, with contributions from Jony Mainaly, Krystel Dossou, Basra Ali, Charles Nyandiga, Lucy Wanjiru, Kathleen Mogelgaard and Assalama Sidi

Gender refers to socially constructed roles, responsibilities and opportunities associated with men and women, as well as hidden power structures that govern the relationships between them. Throughout the world, there are gender-specific differences in consumption patterns, lifestyles, access to and control of resources, power, and vulnerability to environmental problems. Climate change is likely to exacerbate such differences.

There are many well-grounded observations that lend force to the argument that climate change has specific effects on women and men because of the different roles they play in society and their differentiated access to social, economic and physical resources. The Intergovernmental Panel on Climate Change (IPCC) reports that climate change has gender-specific implications in terms of both vulnerability and adaptive capacity (IPCC 2007a). It is generally recognised that women, and especially women living in poverty, bear a disproportionate burden of climate change consequences, and some 70 per cent of the 1.3 billion people in the developing world living below the poverty line are women (Denton 2002). Poor women's limited access to resources, restricted rights, limited mobility and voice in community and household decision-making can make them much more vulnerable than men to the effects of climate change. For example, women predominate in world food production (50–80 per cent) but own less than 10 per cent of land (UNDP 2010a). In Africa, women account for almost 80 per cent of the agricultural sector (Denton 2002) which is a sector particularly vulnerable to the impacts of climate change. Therefore depletion of natural resources and decreasing agricultural productivity are likely to place additional burdens on women (FAO 2005).

In many countries, disparities resulting from the social positions of women within the family and the community may be aggravated by the effects of climate change, making climate change and gender inequality inextricably linked. Climate change slows progress towards gender equality and poses a challenge to poverty reduction efforts, while gender inequality can further worsen the effects of climate change.

Women play a unique role in the stewardship of natural resources, for example by collecting water and fuel. They also provide support to households and communities as active agents of change and shape adaptive mechanisms. Women's responsibilities often involve caring for children who are also vulnerable to climate impacts like extreme weather events. Many argue that women hold the key to adaptation in certain places (Babugura 2010; UNDP 2010a). Women also make an important contribution to disaster reduction, often through participating in disaster management activities and acting as agents of social change (IPCC 2007b). It is therefore vital that gender equality considerations, as well as men's and women's different needs, perspectives, knowledge and skills, be taken into account when planning community-based adaptation (CBA).

This chapter addresses the importance of mainstreaming gender into CBA activities and ways in which gender can be mainstreamed, included or targeted within CBA. Gender mainstreaming is the 'process of assessing the implications for women and men of any planned action, including legislation, policies or programmes, in all areas and at all levels' (Nyandiga and Wanjiru 2011). As well as assessing implications, it is about ensuring that CBA includes women as key actors and beneficiaries. Such gender mainstreaming is the first and arguably the most crucial step towards scaling up CBA projects in a way that is beneficial for both men and women. This chapter therefore focuses on the ways and means to mainstream gender into CBA, on the assumption that once mainstreamed, gender considerations will be more smoothly incorporated into any ensuing efforts to upscale CBA. Effective scaling up of CBA requires the expansion of gender considerations in the climate change agenda in order to relate the situation and vulnerability of women to broader cultural, economic and social processes at the root of climate change processes. Gender mainstreaming is the starting point for this process. As well as explaining why it is always important to consider gender, the chapter explores who the key agents for bringing gender into climate change adaptation are. The case studies outlined below, which were chosen from a variety of different countries and situations, demonstrate how gender can be included as a key element in climate change adaptation activities.

Justifications and evidence

Gender roles and responsibilities differ between regions and localities, and also vary in time, but there is evidence that women are disproportionately impacted by climatic change. Dankelman (2002) notes that women are the main 'energy managers' in many households in the South, for example collecting firewood for fuel. Women in many households also play crucial roles in collecting water and in agricultural production to ensure the family's food security. Overall, poor women live on the frontlines of climate change and have a fundamental role as managers of the environment. They also possess knowledge and experience that is critical for successful responses to climate change (Nyandiga and Wanjiru 2011).

As well as different gender roles, women have differential vulnerabilities. Cannon (2002) argues that there are specific gender attributes which increase women's vulnerabilities in some respects. For example, increased salinity of drinking water has been found to increase maternal health risks during pregnancy in Bangladesh (Khan *et al.* 2011). The UNDP Human Development Report 2007/2008 identified several risk-multipliers that women and other vulnerable populations are particularly affected by, including reduced agricultural production, water scarcity and coastal weather events (UNDP 2007). Nelson *et al.* (2002) therefore argue that the effects of climate change are *very likely* to be gendered as gender inequalities intersect with climate risk and vulnerabilities (UNDP 2007).

Women are also likely to be more vulnerable to the impacts of climate-related disasters. GTZ (2010) finds that, globally, women and children are 14 times more likely to lose their lives in a natural disaster. This may be because the most socially excluded and economically insecure in any community represent those least able to control or access resources for disaster relief and coping mechanisms. For example, during the 1991 cyclone in Bangladesh, 90 per cent of the fatalities were women (UNDP 2010a). In many places women often lack land rights and capacity to access finance, technology, training and information to adapt to climate-related disasters. Lack of land ownership rights causes problems as women may be denied the ability to buy a new plot of land in the aftermath of a disaster. This makes gender-sensitivity crucial for adaptation to impacts and climate disaster risk reduction efforts.

It is also clear that gender is insufficiently mainstreamed in 'many areas of development policy and practice' (Nelson *et al.* 2002) and gender inequalities still persist as a key barrier to development (WDR 2011). There is a clear need to avoid such mistakes when tackling climate change and for the gender element to be recognised in adaptation practice. Women are often absent from decision-making processes at all levels, from the global level (in the United Nations Framework Convention on Climate Change – UNFCCC – and other global processes) to the local level in male-dominated households. Gender mainstreaming has often occurred as an 'afterthought' in development and climate policy (Denton 2002). For example, within the Clean Development Mechanism (CDM) there is no inclusion of gender awareness and the 'large-scale industrial projects' are recognised as providing few benefits for women (UNDP 2010b). Until around the time of the eighth Conference of the Parties (COP 8) to the UNFCCC in New Delhi, India, in 2002 it was argued that gender aspects were generally 'neglected in international climate policy' (FAO 2005) despite the fact that gender, like poverty, is a cross-cutting issue that must be recognised as such. As Rooke (2009) argues; 'there will be no climate justice without gender justice'. Inclusion of gender issues in adaptation efforts at all levels must be recognised as an important equity and vulnerability issue, as well as being crucial to ensure the efficiency of programmes. Successful adaptation will require equal participation because only if all participants' voices are heard, their needs addressed and their knowledge included, will adaptation planning achieve maximum benefits for society (UNDP 2011).

From evidence and concepts to practice

Scaling-up the mainstreaming of gender into CBA from pilots to larger scales will likely involve allocating greater rights and responsibilities to women in adapting to climate change. Focusing on the role of women as critical potential actors in CBA efforts would also help foster necessary changes in approaches to development. Investing in women and girls is already considered a key breakthrough strategy for achieving the Millennium Development Goals (MDGs) at the global level. And there is some evidence that investing in adaptation to climate change that specifically targets women may lower risks from a financial perspective since women are seen as less of a risk financially. This has been evident, for example, in micro-finance projects (Gurung *et al.*, 2009). Thus, incorporating gender may enable CBA to be more financially efficient and effective.

Reality on the ground, however, indicates that development practitioners and policy-makers are often not clear on the 'how' of mainstreaming gender in CBA projects. Gender mainstreaming in CBA projects needs vigorous and sustained participation of both women and men in all project aspects. This will ensure CBA projects have an approach that harmonises sustainable development and efforts to address climate change – through a gender-balanced and culturally appropriate approach. Cultural barriers can be a difficult challenge in implementation of gender-sensitive approaches. There is a need for gender mainstreaming to move beyond the superficial level into actual implementation, which means that key actors must consider this as an inherent part of project design and decision-making rather than as a 'box-ticking' afterthought (FAO 2005).

Hence, gender mainstreaming should integrate gender perspectives into project design and programming – ensuring that the input of both women and men becomes part of the knowledge generated and lessons learned from CBA projects. Gender concerns need to be integrated into CBA projects right from the onset, and throughout the project cycle. Gender mainstreaming is a strategy for making women's experiences and concerns an integral part of design, implementation, monitoring and evaluation of policies, with the goal of achieving gender equality. This will facilitate efficiency in scaling up of CBA projects as it is only when the participation of women is considered that CBA will be truly effective. It will also enable forthcoming CBA projects to contribute to the achievement of gender equality and women's empowerment (UNDP 2010a). Gender mainstreaming has been linked to efficient project development and therefore should receive specific budgets in programmes and projects to ensure that all activities are gender responsive.

Upscaling and mainstreaming gender into adaptation planning: learning by doing

Case study 1: scaling up community-level work in North Kenya, by Basra Ali

This project in Turkana District, North Kenya focused on women's groups as these were often the only organised groups in many parts of North Kenya. This was part of a three-year programme for human security, women's empowerment and livelihood support. Opportunities for scaling up CBA work were found to be among actors, activities and funding processes. With regards to actors it is necessary to ask whether individuals have the space to grow within institutions, to what extent government officers will support them and whether there is competition between actors or the interests of business groups (Ali 2011). Activities must include interventions with a continuous development impact in priority areas for the community. Funding processes and donor parties must move away from discussing their work in board rooms and instead spend time with the specific people that they target.

It was found that women did many of the important and productive work activities in the community, but were disproportionately affected by violence and insecurity. Interventions must consider issues of governance and power, with awareness of tribal politics and religious groupings (Ali 2011). It was recommended that there is a need for home-grown solutions for adapting to climate change. Moreover, there is a need for regional cooperation, new laws to protect women, more local funds (with proper authority and consultation), clearer division of constituencies by donors and government, as well as proper and well-capacitated networks. Only when these issues are addressed will projects such as these be able to 'scale up' into wider interventions.

Case study 2: weathering the storm – girls in a changing climate, by Assalama Sidi

The research by Plan International in 2010 focused on girls aged 13–18 years in drought-prone areas of Ethiopia and flood-prone and cyclone-prone regions of Bangladesh. The study found that gender and age discrimination are two root causes of climate risk. Girls are particularly affected by the impacts of climate change compared to boys due to the roles and responsibilities they are ascribed by society, and because rights to education, health and employment are uncertain. They often spend more time fetching water or firewood during drought and flood periods (Figure 14.1). Moreover, girls are even excluded from activities that reduce risks from floods, such as swimming and tree climbing, which means they are 'less prepared to face the impact of climate change' (Sidi 2011). The research showed that investment in girls is the 'missing piece of the puzzle in terms of long-lasting results in climate change adaptation'. Girls and young women remain ignored in policies aimed at tackling climate change and its

Figure 14.1 Girls in Ethiopia reported that the time they spent fetching water increased substantially each day during drought periods, and that the last few years have seen prolonged periods of drought (source: Plan International 2011).

impacts on development. Equity considerations require that climate change adaptation should focus on those most at risk, and the research showed that prioritising education and participation for girls will result in 'significant benefits' through the empowerment of one of the groups most at risk (Sidi 2011).

Case study 3: gender mainstreaming in Nepal's National Adaptation Programme of Action process (adapted from Mainaly and Chandani, 2012)

Nepal has recently prepared and submitted a National Adaptation Programme of Action (NAPA). UNFCCC NAPA guidelines require each least developed country to incorporate gender elements within the NAPA. Moreover, the NAPA is prepared so as to contribute to the pursuit of national development goals. A study was conducted to explore whether or not Nepal's NAPA and other relevant legal and policy frameworks considered gender. The study also explored whether men and women participated in the formulation of the NAPA and other relevant legal and policy frameworks to advance concerns over gender. The study thus explored both substance (incorporation of gender considerations) and process (participation of women) in the NAPA and other legal and policy formulations in Nepal.

The study by Mainaly and Chandani (2011) found that gender considerations were not addressed in the processes of policy formulation. In terms of substance, however, the NAPA in Nepal was successful at integrating gender up to the level of the vulnerability assessment, but could not adequately address it in planning and implementation frameworks. Studies on agriculture and food security found that declines in agricultural yields cause an increased workload for women, as well as malnourishment. In the arena of public health, since women are more responsible for the care of children, sick and elderly people in the family, health impacts will affect women more than men.

Nepal's development planning framework provides an enabling environment to mainstream climate adaptation programmes into existing development frameworks. Under the Local Self Governance Act (LSGA) of 1996, a law providing a framework for decentralised governance in Nepal, the inclusion of women is enshrined in the planning process as well as targeted beneficiaries of development programmes (Mainaly and Chandani 2011). Thus there is an immense opportunity to mainstream gender considerations into the climate change and development regimes of Nepal since the climate change and development goals of the country are seen as complementary to each other (Mainaly and Chandani 2011).

Case study 4: energy technology transfer to combat deforestation and reduce climate change vulnerability in Zagnanado, Benin, by Krystel Dossou

In Benin, the NAPA highlights a strong link between the domestic energy sector and the forestry sector (Dossou 2011). Deforestation rates are high and women's groups are often most vulnerable to resource depletion due to deforestation. Although deforestation is often linked to climate mitigation, it also impacts on vulnerability since forests can protect against soil erosion, biodiversity loss and water-related impacts of climate change. A CBA project in Benin was thus designed to halve the consumption of fuelwood locally by adopting economic stoves made of local material, and also by promoting tree plantations. Some 300 women were trained to build the economic stoves, and each of these women trained four other women (making a total of 1,200 women). A technical guide on building the fuel-efficient stoves was produced (Dossou 2011). The project encountered several difficulties but also opportunities to learn lessons about ways forward. Challenges included the lack of land for tree-planting, and finding that the stoves could be destroyed by rains, making it necessary to build them under some sort of covered protection.

The project provided a good example of potential synergies between adapting to and mitigating climate change. Experience also suggests that scaling up the project could increase energy savings, reinforce forest resilience and improve human well-being (Dossou 2011). Such scaling up would first need a strong definition of the diffusion mechanism. The replicability of the pilot project would then need to be increased, along with support for the managerial capacities of women. It would also be important to take local differences into account.

A practical toolkit: ways of mainstreaming gender into adaptation projects, by Charles Nyandiga and Lucy Wanjiru

The inclusion of gender equity considerations into development programmes and plans is more often talked about than actually accomplished. While there is consensus on the need to include gender considerations, the practicalities of doing this at different scales are extremely challenging and aggravated by societal norms, institutions and the nature of governments. How, then, can it be ensured that widely agreed intentions to include gender considerations are fulfilled while addressing the challenges and impacts of climate change at local and regional scales and their corresponding impact on poverty in many developing countries? The United Nations Development Programme (UNDP) has developed a guidebook of simple tools and practical advice on how to take a gender-sensitive approach in the planning and implementation of CBA projects (UNDP 2010a) (Figure 14.2).

One key tool is that of 'gender analysis', which is used to identify differences and provide empirical (quantitative and qualitative) evidence for gender roles, activities, needs and for available opportunities for men and women (Nyandiga and Wanjiru 2011). It provides information for gender mainstreaming to ensure projects promote gender equality. It can be applied throughout the project cycle, right from the initiation and vulnerability assessment. Gender-balanced participation is critical in all aspects of project planning and implementation, since projects should be designed to 'accommodate women's traditional roles and responsibilities' (Nyandiga and Wanjiru 2011). If women are not consulted then projects may not be sustainable or appropriate (UNDP 2010a). Overall, a gender-sensitive project is also likely to facilitate effective scaling up of CBA.

The UNDP (2010a) has identified a number of key questions for practitioners to consider when implementing CBA projects to ensure that gender is fully incorporated at all stages of project planning:

- Has relevant gender information, especially socio-economic information, been identified and included in discussions about project formulation?
- Is background data disaggregated by sex?
- Have gender specialists (in the community-based organisation or non-governmental organisation, for example) and representatives of women's organisations within the community been consulted?
- Have both men and women been involved in problem identification?

Innovative approaches: women's empowerment, family planning and resilience, by Kathleen Mogelgaard

Women have unique vulnerabilities and capabilities in the face of climate change, which arise from their roles as primary caretakers of the family and providers of food, water and fuel – the basics of family survival (Mogelgaard 2011). Women's health, well-being and ability to participate in decision-making are intimately connected to pregnancy, child-bearing and child-rearing – women's

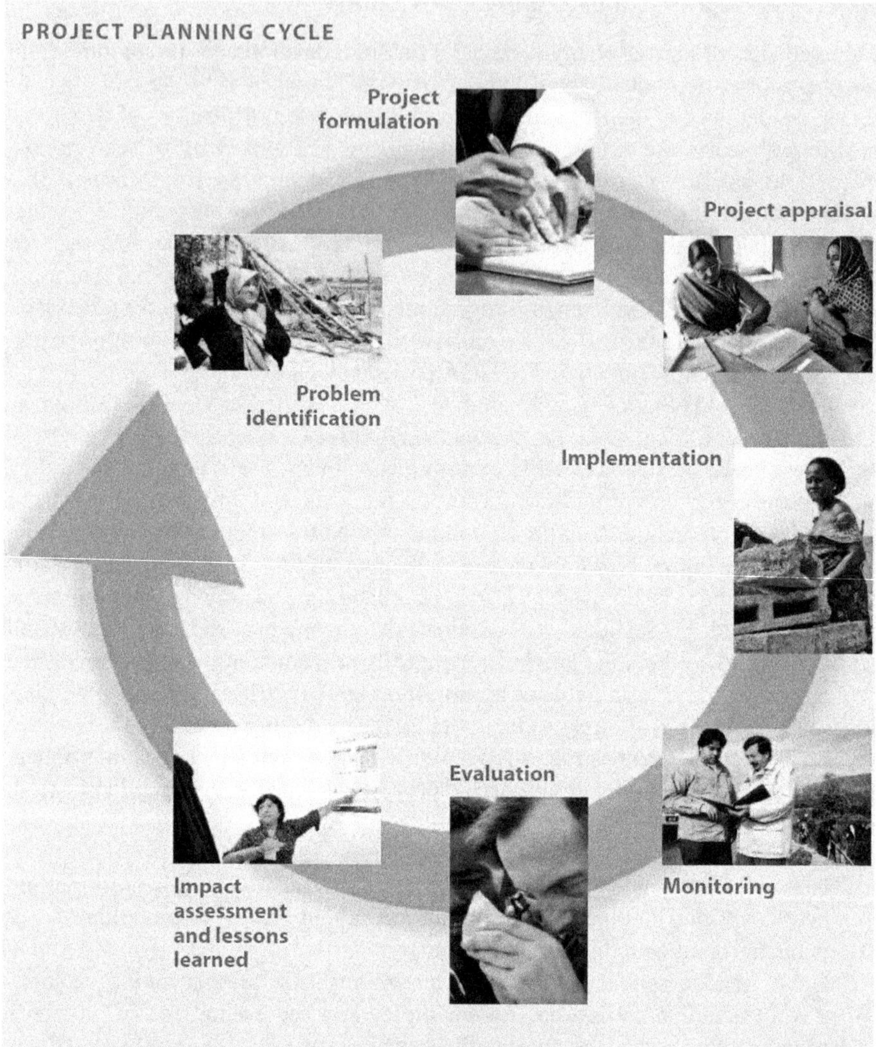

Figure 14.2 CBA project cycle. The figure shows the different stages at which gender considerations should be incorporated in the project cycle (source: UNDP 2010a)

overall reproductive role – and are critical for an effective response to climate change (Mogelgaard 2011). Reproductive health care is embedded as a target in the MDGs with the aim of achieving universal access to reproductive health by 2015. This would reduce maternal mortality, maternal morbidity and unintended pregnancy; it is estimated that 70 per cent of maternal deaths would be averted. Reduction of unintended pregnancies, particularly among adolescents, can improve educational and employment opportunities for women and contribute to improving their status, reducing poverty and spurring economic growth, 'all important foundations for adaptive capacity' (Mogelgaard 2011).

Arguably, meeting the unmet demand for family planning could slow population growth, which could alleviate adaptation challenges and reduce the scale of human vulnerability to climate change at broader scales. More than 215 million women globally have an unmet need for family planning (WHO 2011). Mapping at the global scale has shown that there is a close overlap between areas of high population growth and high vulnerabilities to climate change (Mogelgaard 2011). There are lessons to be learned from population–health–environment (PHE) projects worldwide that provide tremendous opportunities for strengthening CBA approaches. In some instances families may choose to have many children as a response to high mortality rates and in order to support their family. Having many children may help offset risks and reduce vulnerability in some cases. Empowering women and providing them with improved health care overall is likely to reduce this need. Overall, a rights-based approach providing universal access to women's health services is a no-regret, win–win opportunity when it comes to climate change adaptation.

Conclusions

There is an urgent need to align climate change adaptation efforts with Millennium Development Goal 3 in order to 'promote gender equality and empower women'. Lack of success in achieving gender equity and female education targets is likely to hinder progress in adapting to climate change impacts. The case study by Plan International shows that women often have a different risk profile under climatic events (Sidi 2011). Moreover, since women are often involved in caring for children, targeting women in CBA projects may help improve resilience among their children, who will also be disproportionately impacted by future climate change. Thus, in addition to improving intra-generational gender equity, recognising the role of women potentially improves inter-generational equity. This case study highlights that education for girls and young women will play a crucial role in tackling the root causes of climate risk (Sidi 2011). Provision of services for women, including family planning (Mogelgaard 2011), can impact on women's ability to adapt so there is need for women's health and maternal health to be considered as an opportunity for reducing vulnerability. Although PHE projects have not been designed with climate change in mind, they provide an opportunity for holistic win–win situations to reduce climate risk while improving the status of women. Overall, FAO (2005) argues that targeting women and the introduction of a gender-sensitive approach 'could assure real benefits both in efficiency and equity terms'. This demonstrates the importance of efforts to scale-up and mainstream gender-sensitive CBA.

Scaling up CBA will occur at the local, national and international levels. In their study of Nepal's NAPA, Mainaly and Chandani (2011) recognise that there are great opportunities to mainstream gender at the national policy level in addition to the local level, because while gender is being successfully integrated into vulnerability assessments, it is not yet adequately addressed in higher levels of planning and implementation. At the international level, emerging global funds for adaptation such as the Adaptation Fund should fully integrate gender equality considerations into operational policies and guidelines, as well as project review criteria, to ensure that the needs and interests of women are well represented (UNDP 2010c). Climate change adaptation efforts must align with achieving priorities such as the MDGs on gender equity (Goal 3) and access to reproductive health (Goal 5) at the global level.

The mainstreaming of gender considerations in CBA projects is an essential prerequisite for scaling up such projects. The case studies show that scaling up gender-sensitive CBA from pilots to larger scales may involve allocating greater rights and responsibilities to women. In terms of the 'how to' of mainstreaming gender into CBA, the UNDP has developed a guidebook of practical advice on how to take a gender-sensitive approach in CBA projects (UNDP 2010a; Nyandiga and Wanjiru 2011). Practitioners must ask themselves whether women's experiences and concerns have been an integral part of the whole project cycle (Figure 14.2). The monitoring and outputs of projects should also be tested for gender impacts, for example using the UNDP's 'Gender Mainstreaming Scorecard' (OECD 2011). Similar efforts must be applied to CBA for programmes and projects to be effective. Since women's groups are well organised in many places (Ali 2011), capacity building and expanding these networks of cooperation provide an opportunity for scaling up CBA. Dossou (2011) also suggests that we must explore the diffusion, replicability and reproducibility of such pilot projects in order to facilitate the scaling up of CBA.

Parallels can be drawn from the field of forest conservation, where efforts towards empowering women have aided the long-term sustainability and efficient functioning of projects (Gurung *et al.* 2009). In the final case study, Krystel Dossou provides the example of fuel-efficient stoves in Benin as an opportunity to improve human well-being and reduce the deforestation which exacerbates local climate vulnerability (Dossou 2011). There are efforts geared towards scaling up and mainstreaming gender into mechanisms developed under the UNFCCC to reduce greenhouse gas emissions from deforestation and forest degradation under REDD (Reducing Emissions from Deforestation and Forest Degradation) projects. Gender-mainstreaming tools such as gender analysis, gender-differentiated data and gender-sensitive participatory tools have been used in such forest projects to assess and include women's perspectives with positive results (Gurung *et al.* 2009).

Overall, the key agents of change will be national and sub-national policymakers, development planners, community leaders, women's groups and women in vulnerable communities themselves. Successful gender mainstreaming will integrate gender perspectives into project design and programming, to ensure that the input of both women and men become part of the knowledge generated and the lessons learned from CBA projects (Nyandiga and Wanjiru 2011). Considering the needs and inclusion of women will be crucial to facilitate effective scaling up of CBA projects that benefit men, women and children. It will also enable emerging CBA projects to contribute to women's empowerment and achieve gender equality. Wider co-benefits can be achieved by focusing on women, for example in the case study of fuel-efficient stoves which conserve energy, reduce deforestation as a driver of vulnerability and improve health by reducing indoor emissions (Dossou 2011). In summary, women must be recognised as pivotal agents in climate and development processes in order for climate change adaptation efforts to be both effective and equitable.

References

Ali, Basra, 2011. Climate change adaptation: scaling up intervention programmes. The Case of Northern Kenya Women Empowerment and Livelihood Support Programme: 3 year programme. Presentation at the 5th Annual Community Based Adaptation to Climate Change Conference, Dhaka, Bangladesh, 28 March.

Babugura, Agnes, 2010. Gender and climate change: South Africa case study. Heinrich Boll Stiflung, The Green Political Foundation. [Online] Available at: www.boell.de/ecology/climate/climate-energy-South-Africa-9074.html (accessed 19 September 2013)

Cannon, Terry, 2002. Gender and climate hazards in Bangladesh. *Gender and Development*, Volume 10, No. 2.

Dankelman, Irene, 2002. Climate Change: Learning from gender analysis and women's experiences of organising for sustainable development. In *Gender, Development and Climate Change*, Ed. Rachel Masika, Oxfam Publishing.

Denton, Fatma, 2002. Climate change vulnerability, impacts and adaptation: Why does gender matter? *Gender and Development*, Volume 10, No 2.

Dossou, K., 2011. Energy technology transfer to combat deforestation and reduce climate change vulnerability in Zagnanado (Benin, West Africa). Presentation at the 5th Annual Community Based Adaptation to Climate Change Conference, Dhaka, Bangladesh, 28 March.

FAO, 2005. *Gender: The Missing Component in the Response to Climate Change*. Food and Agriculture Organisation of the United Nations.

GTZ, 2010. *Climate Change and Gender: Economic Empowerment of Women Through Climate Change Mitigation and Adaptation*? [Online] Available at: http://cedoin-gtz.padep.org.bo/upload/cambio-climatico-genero.pdf (accessed 19 September 2013).

Guring *et al.*, 2009. *Engendering REDD Workshop*. IUCN, GGCA and WOCAN. [Online] Available at: www.generoyambiente.org/arcangel2/documentos/499.pdf.

IPCC, 2007a. Assessment of adaptation practices, options, constraints and capacity. In *Climate Change 2007: Impacts, Adaptation and Vulnerability. Contribution of Working Group II to the Fourth Assessment Report of the Intergovernmental Panel on Climate Change*, M.L. Parry, O.F. Canziani, J.P. Palutikof, P.J. van der Linden and C.E. Hanson, Eds., Cambridge University Press.

IPCC, 2007b. Human health. *Climate Change 2007: Impacts, Adaptation and Vulnerability. Contribution of Working Group II to the Fourth Assessment Report of the Intergovernmental Panel on Climate Change*, M.L. Parry, O.F. Canziani, J.P. Palutikof, P.J. van der Linden and C.E. Hanson, Eds., Cambridge University Press.

Khan, A., Ireson, A., Kovats, S., Mojumder, S., Khusru, A. and Rahman, A. 2011. Drinking water salinity and maternal health in coastal Bangladesh: Implications of climate change. *Environmental Health Perspectives*, Volume 119, No. 9.

Mainaly, J. and A. Chandani, 2011. A working paper on climate change laws, policies and programs: An opportunity for gender mainstreaming in Nepal. Presentation at the 5th Annual Community Based Adaptation to Climate Change Conference, Dhaka, Bangladesh, 28 March.

Mogelgaard, Kathleen, 2011. Healthy women, healthy planet: Women's empowerment, family planning, and resilience. Presentation at the 5th Annual Community Based Adaptation to Climate Change Conference, Dhaka, Bangladesh, 28 March.

Nelson, V., Meadows, K., Cannon, T., Morton, J. and Martin, A., 2002. Uncertain predictions, invisible impacts, and the need to mainstream gender in climate change adaptation. *Gender and Development*, Volume 10, No 2.

Nyandiga, Charles and Wanjiru, Lucy, 2011. How to mainstream gender into CBA. Presentation at the 5th Annual Community Based Adaptation to Climate Change Conference, Dhaka, Bangladesh, 28 March.

OECD, 2011. *Gender Indicators: What, Why and How*? [Online] www.oecd.org/dataoecd/46/47/43041409.pdf (accessed 19 September 2013).

Plan International, 2009. Weathering the storm: Adolescent girls and climate change. [Online] Available at: www.plan-uk.org/resources/documents/35316/ (accessed 19 September 2013).

Rooke, A., 2009. *Doubling the damage: World Bank Climate Investment Funds Undermine Climate and Gender Justice*. [Online] Available at: www.genderaction.org/images/2009.02_Doubling%20Damage_AR.pdf (accessed 19 September 2013).

Sidi, Assalama, 2011. Weathering the storm: girls and climate change. Plan International. Presentation at the 5th Annual Community Based Adaptation to Climate Change Conference, Dhaka, Bangladesh, 28 March.

UNDP, 2007. *UNDP Human Development Report, 2007/2008*. UNDP.

UNDP, 2010a. *Gender, Climate Change and Community-Based Adaptation: A Guidebook for Designing and Implementing Gender-Sensitive Community-Based Adaptation Programmes and Projects*. UNDP.

UNDP, 2010b. *Clean Development Mechanism: Exploring the Gender Dimensions of Climate Finance Mechanisms*. [Online] Available at: www.beta.undp.org/content/dam/undp/library/gender/Gender%20and%20Environment/Gender%20Dimensions%20of%20the%20Clean%20Development%20Mechanism.pdf (accessed 19 September 2013).

UNDP, 2010c. *Adaptation Fund: Exploring the Gender Dimensions of Climate Finance Mechanisms*. [Online] Available at: www.adaptationlearning.net/sites/default/files/Adaptation%20Fund%20final%202010.pdf (accessed 19 September 2013).

UNDP, 2011. Africa Adaptation Programme experiences. gender and climate change: Advancing development through an integrated gender perspective. Discussion Paper Series, Vol. 1.

WDR, 2011. *The World Development Report 2012: Gender Equality and Development*. WDR.

WHO, 2011. *Global Health Observatory: Universal Access to Reproductive Health*. [Online] Available at: www.who.int/gho/maternal_health/reproductive_health/en (accessed 19 September 2013).

15 When disaster is an opportunity

Community-based disaster response as a tool for scaling up risk management

Norberto Carcellar, Jason Rayos Co, Zarina O. Hipolito, Diane Archer and Somsook Boonyabancha

When a large disaster hits – like the 2004 Indian Ocean tsunami – it receives international media coverage, aid is mobilised and aid agencies rush to respond. While survivors of smaller disasters might wish for such attention, there are often serious negative side-effects to these responses. As in development, the effectiveness of external funding depends on the quality and orientation of the intermediary (mostly local) institutions through which funding is channelled. Yet often, these institutions do not facilitate the participation of survivors in the process. Survivors may be sidelined with little influence over the responses that are chosen by more powerful groups, and little control over how the external funding is used or prioritised. For example, the disaster response in Banda Aceh after the 2004 Indian Ocean tsunami hit was referred to as the 'second tsunami' by the local communities there (Syukrizal *et al.* 2009). Among the many international agencies, each had their own budgets and priorities – and survivors found it very difficult to influence what was done and to get their priorities supported. While actors may employ the rhetoric of participation, for example in post-disaster housing reconstruction, the reality of this can range from local communities simply providing labour, to taking an active role in project management and planning (Davidson *et al.* 2007). Donor-driven recovery and reconstruction processes may focus on producing visible outputs, while the humanitarian cluster system seeks to divide problems into manageable pieces, but this is at the cost of addressing the bigger picture, the underlying vulnerabilities and the priorities of the disaster survivors (Levine *et al.* 2012; Schilderman and Lyons 2011; Schilderman and Lyons 2010; Lyons 2009).

This can be particularly challenging in the urban context of multiple actors and complex governance structures (Zetter and Deikun 2010), as experience in Haiti has shown (Pantuliano *et al.* 2012; Levine *et al.* 2012; ALNAP 2012).

Disasters always bring tragedy, but they also open up an opportunity for change in the affected communities. The post-disaster recovery period offers a chance to turn a negative and desperate situation into a possible longer-term positive outcome. Having a clear understanding of the opportunities that arise as a result of a disaster and how to make the most of them through the rebuilding process leads to a greater ability to provide future support and prevention. Post-disaster situations are a vital opportunity to incorporate disaster risk reduction practices for building future resilience (ALNAP 2012; IPCC 2012). While this is a challenge, it is also an opportunity – to develop ways of working with the existing institutional framework of local government and civil society organisations (Zetter and Deikun 2010). For aid agencies and local institutions to provide the appropriate assistance, it is necessary to understand how the disaster can be viewed as an opportunity to make positive structural change.

The participation of local vulnerable groups in decisions about large-scale disaster response is critical for realising this opportunity to ensure that these responses address vulnerability priorities, contribute to long-term resilience and are implemented in sustainable and scalable ways (Schilderman and Lyons 2011; ALNAP 2012; UNISDR 2009). Low-income and already disadvantaged groups are the most vulnerable to disasters.[1] Their rights, which are already weak, will be further weakened as a result, pushing them even further to the margins of society. Sometimes, disasters can push previously not-so-poor people into a lower-income category. If these disaster victims were not organised as a collective before the disaster, they may lack the necessary social linkages to help them recover from the catastrophe. Therefore, the response to the disaster should be to provide not only short-term relief, but also the necessary assistance and tools so that longer-term change and improvement can be achieved by the people themselves.

So how can the role of vulnerable communities be 'scaled up' to ensure community needs and priorities are reflected in large-scale disaster response, and what kinds of institutions are needed? This chapter describes initiatives in the Philippines that have enabled local communities to take the lead in rebuilding their lives and neighbourhoods. It has two key messages. First, the survivors of disasters should be looked at in a new way, not simply as helpless and dependent victims; rather, they should be regarded as agents for change in rebuilding their lives and their communities. Second, with the right knowledge and techniques (and the right local community capacity), outsiders can help the survivors to harness their energy positively and to empower themselves through the stages of emergency relief and rehabilitation. This 'how to' aspect is where development agencies often stumble – they struggle with the design and use of the support tools that will create a platform to allow this energy to be developed and used. There is room for improvement in international approaches that intervene in and support the change process in affected communities, and deal with the added complexity of disasters in urban areas (Zetter and Deikun 2010). With such knowledge, development agencies and local institutions will be better equipped to respond to climate change and support community initiatives for adaptation on a larger scale.

The Homeless People's Federation Philippines, Inc[2]

Carcellar *et al.* (2011) describe the experience of the 'Homeless People's Federation of the Philippines Inc.' (HPFPI) in addressing disaster risk reduction (DRR) through community-rooted interventions in the Philippines. The experience of the

HPFPI in DRR has evolved from years of post-disaster response to poor communities that have been hit by calamities (Rayos Co 2010). As a learning organisation, the federation has sharpened its disaster-related capacities with every community-driven intervention it has supported. The federation has worked closely with its support NGO, the Philippine Action for Community-led Shelter Initiatives, Inc. (PACSII),[3] to develop its support for disaster-affected communities to encourage community-rooted and driven post-disaster response. These two agencies are referred to as the Philippine Alliance. Its approach includes identifying the ways in which government and other stakeholders can support the needs and address the vulnerabilities of at-risk communities, mapping out possible entry points for incorporating DRR into the effort to scale-up community-driven initiatives to the city level. It is a member of the Shack Slum Dwellers' Federation, and has drawn on these methodologies to sharpen its own organising practice.

The Philippines maintains a high disaster-risk profile, attributable to its location within the Circum-Pacific belt (Pacific Ring of Fire) and the typhoon belt of the North Pacific basin. Consequently, the Philippines suffers from frequent earthquakes, volcanic eruptions, lava and mudflows, landslides, typhoons, floods and droughts that adversely impact on the populace, most especially the poor. This constitutes a major development issue as the frequent disasters, natural or human induced, hinder government efforts to reduce poverty and lessen vulnerabilities of many low-income, hazard-prone-located communities. Furthermore, climate change has probably exacerbated the intensity of recent extreme weather events – for instance during the September–October 2009 mega floods caused by strong winds and extremely high levels of rainfall (100–450 millimetres within the first 24 hours) occasioned by typhoons Ketsana (Ondoy) and Peping (Parma), which resulted in 1,212 fatalities, countless displacements and up to Php 38 billion ($852 million) in total damages (Carcellar *et al.* 2011). Climate change is also likely to continue exacerbating the intensity and/or frequency of these and other extreme weather events.

In 2000, the 'Payatas trash slide' was triggered by heavy rains, spearheading the involvement of the HPFPI and PACSII in disaster response. The tragedy took the lives of 288 people, buried hundreds of homes and displaced close to 1,000 waste pickers (Carcellar *et al.* 2011). With no background in disaster response, the HPFPI engaged in rescue and relief coordination and later supported the trash slide survivors to acquire land and re-build their houses away from the garbage dump. The experience in Payatas provided a wealth of learning on disaster-related responses that would help improve later interventions in other disaster-affected communities, as in the years following the trash slide the Alliance had time to reflect on and assess the long-term effects of its interventions on the lives of the community.

From 2006 onwards, the Alliance had its hands full in this regard, with disaster situations involving thousands of marginalised families. In 2006, the HPFPI was involved in two major disasters, namely the landslide that buried the whole *barangay* of Guinsaugon in Leyte province, and the Mt Mayon mudflow and subsequent flash flood in Bicol Region. Combined, these disasters killed at least 362 people and thousands more were recorded missing. They also caused massive devastation to livelihoods, infrastructure and agriculture. In recent years the Alliance has supported communities that have been the victims of fire and flash floods. In 2007, 246 houses were razed by fire in Lower Tipolo *barangay* in Mandaue City and almost 1,000 people were left homeless. In 2008 Typhoon Frank brought 354 millimetres of rain within a 24-hour period, flooding 180 villages in the city of Iloilo; up to 500 people were killed and 261,335 were affected (Carcellar *et al.* 2011). In 2009 at least 464 people died, 529 were injured and hundreds of thousands left homeless by the widespread flash flood caused by Typhoon Ketsana in Metro Manila and neighbouring provinces (Carcellar *et al.* 2011).

The successive onslaught of disasters in recent years has highlighted the multiple vulnerabilities of low-income communities in times of natural or even human-induced calamities. But likewise, these disasters have brought out the inherent strength of communities to rise above the tragedies and harness their experience to create local learning, which can be used in designing community-driven DRR interventions. The thrust of the Alliance – to address the increased vulnerabilities of communities with appropriate support mechanisms – fuels and sustains the community-driven initiatives in disaster response and risk reduction.

Disaster responses rooted in communities

As a result of dealing with these disasters and fine-tuning subsequent interventions, the HPFPI has developed a post-disaster response process that is rooted in communities and driven by their articulated needs.

Initial data gathering

The post-disaster response team's initial activity at ground zero is to gather data on the severity and scope of destruction in terms of lost lives, numbers of affected families and immediate needs of victims. During this stage, the HPFPI also touches base with disaster response structures such as the Disaster Coordinating Council. The data-gathering process is used both as a disaster documentation tool and an entry point strategy for community and local government engagement (Satterthwaite 2011). The information gathered is used to assess the immediate needs of families, especially women, children and the elderly, for inclusion in the appeal for emergency funding to international organisations involved with the HPFPI network.

Since HPFPI volunteers are rooted in communities,[4] they are often the first to undertake damage assessment of the aftermath of disaster, and the initial information they produce is what local government units (LGUs) and outside organisations involved in rescue and relief operations rely on immediately after tragedy strikes. For example, in the massive flooding that submerged almost all of Metro Manila as a result of the heavy rains caused by Typhoon Ketsana in

September 2009, HPFPI volunteers, who themselves were disaster victims, rapidly provided accurate data to the authorities and aid organisations concerning the scope and gravity of destruction that the flooding had brought to their communities. Local community networks can play an essential role in needs assessment and targeting (ALNAP 2012).

Trust and contact building

In the initial stages of intervention, HPFPI leaders expect scepticism and an outright refusal to cooperate from disaster-struck communities, and even local authorities, regarding their programme. Federation leaders thus invest in trust and contact building by introducing the HPFPI and its interventions to the village authorities, local leaders and also to city or municipal government. If there is not already a presence within the community, experienced volunteers consider this process as among the most arduous challenges in the disaster response cycle. There is much insecurity and the process of establishing and legitimising the HPPFI presence in the community can create foes as well as allies. Ultimately, however, it helps cement the relationship between the community and the HPFPI. Another effective strategy for trust and contact building is learning exchanges and exposure visits to other communities, which adopted the HPFPI programme and were given necessary support by the Alliance to implement their post-disaster rehabilitation and reconstruction plan. Once local community leaders and members have a chance to understand how the federation process has helped develop a community-led plan for recovery, then there is greater support.

Savings programme implementation

At the heart of all HPFPI intervention is the savings programme. This can be difficult to introduce to disaster victims who often have lost much, if not all, of their material possessions as a result of the tragedy, and generally feel that they should be at the receiving end of resource generation and not the initiators. The experience of community leaders who introduced a savings programme to the mudflow and flash flood victims in Bicol Region best illustrates this point. They were accused by local government officials and some non-government organisations (NGOs) operating in the area of extorting money from the victims. Likewise, community people did not believe that a few pesos each week could be enough to finance a land acquisition scheme. The accusations cast doubts on the effectiveness and credibility of savings to finance a long-term reconstruction plan. But the time and hard work invested in trust and contact building countered people's negative perceptions about savings and it paid off in the long run. The community's savings enabled them to purchase land within only a year of implementing the programme, thereby demonstrating that savings mobilisation can be a vital instrument in disaster response.

Organisational formation and registration

Alongside the savings programme implementation is social preparation for the formation and registration of a community association, as part of the organising work. This part of the disaster intervention process usually takes a couple of years to come to fruition as it entails a lot of preparation on the ground. While working on increasing the savings funds, the HPFPI and the communities also work on capacitating local volunteers and potential leaders to take on lead roles in mobilising people. Once the community organisation is formally created and registered, the HPFPI then assists in its development and strengthening.

Intervention identification

Disaster-affected communities choose the kind of interventions they prefer depending on their needs. To date, the Alliance has developed three types of disaster intervention: an immediate housing materials loan; transit housing for the medium term where this is possible; and a longer-term approach for the acquisition of alternative safe land followed by housing construction.

Consolidating interventions

The federation process seeks to directly address feelings of helplessness and vulnerability. The absence of financial resources can create major obstacles in the daily struggle of low-income households even in normal circumstances, since in the city access to everything from food, education, health and mobility depends on the financial resources of the household. For this reason, savings are helpful to enable households to begin to accumulate the basis for reduced vulnerability. The federation have long experience of saving and lending for multiple purposes and they share their expertise with new groups as they are formed. As groups are established then possibilities for some housing improvements emerge. As noted above, these vary over time and depend in part of the context that people face. Some households face particular vulnerabilities because of the location of their homes and the risks attached to locations under bridges, along coastlines and near open dumpsites, catchment basins and landslide-prone areas.

At the centre of the response to vulnerability is a stronger organisation within communities so that collective action can move the community to plan a collective development agenda and act towards achieving it. The organisational strategy includes action at three levels:

- *Local organisational formation and development* within the neighbourhood and settlement, strengthened by savings practice.
- *Urban poor networking* allows for sectoral planning and strategy development among urban poor organisations for engagement at the city level. This steps up the organising work to include other urban poor federations working towards the same sectoral goals within a city or municipality and

also promotes the same principle of community empowerment and leadership in their development agenda.

- *LGU collaboration with the federation and networks* opens the door for support from the LGU, which is the main player in the city development process because it has control over public resources and policy formulation at the city level. Recognising that there are limits to what community organisations can accomplish alone to achieve risk reduction (Hardoy *et al.* 2001; Dodman and Mitlin 2011), such collaborations are essential to ensure that national and local policies and development decisions support community-driven actions (Satterthwaite 2011).

Scaling up risk reduction

The HPFPI is slowly but deliberately incorporating initiatives for risk reduction in addition to its disaster response initiatives as part of its avowed mission to organise and mobilise low-income communities in high-risk locations. At the heart of the federation's efforts on DRR is the identification of highly at-risk communities in cities and municipalities where the HPFPI is active.

The citywide enumeration (Patel *et al.* 2012) is the main tool used by the HPFPI to identify and support at-risk communities. Community surveys began after the Payatas trash slide catastrophe in 2000. Federation leaders reflected on the importance of accurate and updated documentation on informal settlement after the official tally of deaths and missing persons reported by the government and the media did not even come close to the number of lost lives estimated by community people themselves, who were familiar with the vicinity and the people living and working there. The lack of government documentation on low-income communities, many of them located in high-risk areas, pushes marginalised people further into invisibility. Although informal settlements are visible just about anywhere within the city proper, in the Philippines there is little or no record of their existence other than what is provided by the National Statistics Office (NSO) in the general population census. Because of this, their needs are not included in city development and DRR planning.

The Alliance is now in the process of identifying and profiling at-risk communities in 12 cities and ten municipalities located in Luzon, Visayas and Mindanao. Budget constraints, however, have necessitated the prioritisation of communities that face the greatest risk. Results of the initiative will give an accurate profile of marginalised settlements that are highly at risk of disasters at both the regional and city levels, thereby raising awareness among community members of the dangers of living in unsafe environments. Two further benefits of this process are the chance to build community-based information systems on informal settlements on a city and municipal scale; and to use enumeration results as inputs for community-driven initiatives, local government planning and network building.

The Alliance has likewise partnered with an institution that is knowledgeable in disaster mapping and risk assessment. This reflects awareness that, although the communities are knowledgeable about their local situation, they do not understand all of the risks they face. Environmental Science for Social Change (ESSC) is helping the Alliance to capacitate communities in disaster risk assessment and mitigation by designing a popularised module that will enable a community-driven risk assessment and mitigation (CRAM) process to be implemented. The federation overlays the CRAM results with actual local maps that ESSC helped develop for the HPFPI's selected high-risk communities. The enumeration, CRAM results and map overlays combined will provide a two-dimensional profile that shows both the socio-economic and environmental vulnerabilities of communities to disaster. This profile is a powerful tool that can bolster the credibility of the HPFPI in engaging with city stakeholders, especially the LGUs, on the matter of DRR planning, which includes the issue of marginalised communities.

The limited resources of the federation cannot sustain the regular and comprehensive conduct of enumerations and map overlays in high-risk communities. The HPFPI is thus advocating that local government at *barangay* and city levels incorporate enumeration and risk assessment in their development planning so that more marginalised communities can be profiled and assessed. The results of these enumerations should be taken into consideration by both the *barangay* and the city government in their climate change and disaster response and risk reduction plans, as called for in the newly enacted Climate Change Act 2009 (Republic Act 9729) and the Philippine Risk Reduction and Management Act 2010 (Republic Act 10121).

DRR is likewise being incorporated into efforts to scale-up community development initiatives at the city level. The scaling up effort will be pushed forward, supported and sustained by four factors: strong community networks, an alternative financial facility, a technical support institution and a community-based information system, which are currently at different stages of development. The alternative finance facility, known as the Philippine Action for Slum Upgrading Facility, Inc (PASUFI) will provide loans to the organised urban poor and will help to financially support the scaling up process. It is a potential vehicle for risk sharing as it can include among its products disaster insurance for the poorest families. Reconstruction or rehabilitation loans can likewise be included in its potential list of financial products that cater to highly at-risk communities. PASUFI has been registered as an NGO-MFI (micro-finance institution) which will support the scaling up process, particularly focusing on high-risk communities.

The technical support institution, registered as the Technical Assistance Movement for People and Environment, Inc. (TAMPEI), will contribute to risk mitigation measures by promoting and encouraging the use of green architecture in the houses that participants will design. By taking into account the geography and topography of the environment where the houses will be built, the design can combine maximum efficiency in house use with minimum exposure and susceptibility to disaster. The same principles will hold true whether communities are working on the design of small- or large-scale upgrading initiatives, and will be incorporated when community development initiatives are scaled up citywide.

The last pillar of the scaling up process is the community-managed information system, which will be the primary source of data that can help identify communities in dangerous areas and the risks they are facing. This system is still being designed to the advantage of the communities who will be the end users.

While the efforts of communities to develop and implement their own disaster-related interventions, moving from response to risk reduction, have produced concrete results, much still needs to be done to move these development initiatives from small-scale success stories to a mainstream approach to city development. Communities can only maximise and control the development drivers that are inherently theirs, such as their savings, acquired capacities and the support given by the HPFPI and PACSII. Other development drivers such as land provision, the formulation of enabling policies at the city and national levels and institutionalisation of the community-driven process at *barangay* and city levels are among the supports that city stakeholders can provide to match the efforts of communities.

The communities' limitations largely involve resources and the legal framework within which they have to operate. With external support, however, community-driven DRR holds many possibilities in terms of appropriately addressing the core vulnerabilities of low-income communities to calamities, from improved organisational capacities through to developing emergency response and recovery mechanisms, thus building their resilience as communities (Twigg 2007).

Box 15.1 Iloilo City

The interplay of institutional partnerships of communities with local government, academia, private organisations and international partners is exemplified in Iloilo, where the post-disaster response initiative has been scaled up to the city level. The local government recognised that the urban poor sector and their support organisations are partners in the city's development. It has opened opportunities/venues for the sector to participate in local decision-making through representation in technical working groups and multi-sectoral bodies. Delivery of housing, upgrading, post-disaster assistance and other basic services became massive in scope because of resource sharing that resulted from the partnership built between the LGU and the communities through the Alliance.

The partnerships established with local and national governments and grassroots organisations started long before Typhoon Frank caused devastation in the city. The active participation of the Alliance in technical and multi-sectoral working groups, made possible through executive orders and other local policy structures, opened a wider space for engagement to effect change in local policies. As a member of the Resettlement and Monitoring Task Force, the Alliance assisted in social preparations and other local government resettlement-related activities. This included an information dissemination campaign among communities living in danger zones (along riverbanks, shorelines and those directly affected by the city's infrastructure projects), who were to be transferred to government relocation sites.

Extending facilities and equipment, and detailing personnel (site engineers, surveyors, mappers) to provide technical assistance to housing and disaster rehabilitation measures/interventions of the Alliance were the major contributions of the local government. This kind of assistance effectively lowered the cost of projects managed by the Alliance. The local government formally allocated a portion of the 16.2 hectares of the San Isidro relocation site (in the Jaro district of Iloilo) to those in need who were working with the federation, and also provided for the construction of temporary housing units and communal facilities for Typhoon Frank-affected families, and formalised and facilitated counterparting and collaboration.

The city government, through the Iloilo City Urban Poor Affairs Office (ICUPAO), assisted in the social mobilisation activities of the federation, which included mapping of high-risk/disaster-affected communities, and identification and prioritisation of communities to be given post-disaster assistance (temporary houses and materials loan assistance for housing repairs). The communities have linked with various academic institutions (University of San Agustin–College of Engineering and Architecture; University of the Philippines–Visayas; and Western Visayas College of Science and Technology), which provide technical support to poor communities living in high-risk areas. This support is in the form of documentation and research on alternative building technology, engineering and architectural or logistical services.

Central to the Iloilo experience in multi-stakeholder and scaled up disaster response and DRR intervention is the quality of organisational and governance strength of each of its urban poor communities, which allows them to act on their development agenda. A key factor in obtaining empowered communities is the confluence of determined and enabling community leaders able to mobilise their members into undertaking collective visioning, planning and action, and the unwavering support of said members derived from a keen understanding of such a vision. Good community leadership, capable of multi-tasking, has contributed to the efficiency and effectiveness of communities in managing local and citywide development initiatives, including post-disaster rehabilitation measures. Specific to the experience of the Alliance in Iloilo, community associations together with the Iloilo City Urban Poor Network (ICUPN) are used to help identify and prioritise disaster-affected families as target beneficiaries of temporary housing and materials assistance programmes.

An important realisation emerging from this experience is the need to leverage additional local resources, particularly land for transit and (permanent post-disaster) housing projects to be replicated and scaled up. Cities can access national calamity funds as well as their own calamity funds, which amount to 5 per cent of their total budgets; this can be used for both post-disaster response and pre-disaster risk reduction activities.

Conclusion

Post-disaster responses can be a positive opportunity for change if survivors are the agents for change at all stages, from immediate relief to planning, reconstruction and implementing effective disaster prevention processes for the future. The benefits are clear:

- Community-led responses ultimately prove more effective and long-lasting, as those who are affected know best what their needs and priorities are; properly supported, survivors can end up in a better position to address underlying vulnerabilities, particularly if they are able to negotiate land tenure and build better homes, thus ensuring that 'the most effective adaptation and disaster risk reduction actions are those that offer development benefits in the relatively near term, as well as reductions in vulnerability over the longer-term' (IPCC 2012: 20).
- The contribution of community-led DRR and response can be enhanced with the recognition and support of local government and agencies (UNISDR 2009), leading to scaled up action and wider political and structural changes, by encouraging local authorities to invest further in protective infrastructure or allocate land for vulnerable, low-income groups (Dodman and Mitlin 2011; Satterthwaite 2011).

The appropriate response in a post-disaster situation is not simply the provision of financial and physical resources.[5] It is also a question of unlocking and organising the energy of the survivors so that they can rebuild their lives together. A collectivity of communities facing a similar situation will have a lot of potential to achieve change by making use of their power, and in a post-disaster situation this potential will be heightened by need. Even with only very limited funds, the collective energy can make the financial resources that are provided as relief go much further. The resources can reach beyond rebuilding housing, to chicken farming, boat building, organising a community development fund or addressing other needs that people have thought about in rebuilding their lives. Giving survivors the capacity to manage their communal needs, development and rehabilitation through the provision of flexible financial resources will gradually release their energy, which is amplified by their need to survive following the disaster. This also provides an opportunity to rebuild social networks which may have been weakened by the disaster (Schilderman and Lyons 2010).

The most important role that relief and development agencies can play in a post-disaster situation is to understand the importance of creating a space where the affected people can come together to instigate change. They need a platform where they can link up with other similarly affected groups, in order to rebuild their lives and their communities as soon as possible, with secure livelihoods, and where they can re-establish their rights and form new relationships within the local system. A post-disaster situation encompasses many issues, not just the question of rebuilding houses – it also brings up the question of human rights, and social and political relationship change. Livelihoods need to be revived, and local communities need to be re-established.

This opportunity can be seized right from the earliest stages of relief efforts, in the relief camps. If the affected persons have a chance to talk and discuss with each other, as is possible in a relief camp where all the victims are re-grouped, then they can think together and express their ideas about what they want to do to recover from the disaster. The discussion process itself is vital, as through it they can form a belief in what they want to do and become increasingly confident that they can achieve it themselves – that they are the agents of change. As relationships begin to form, things will start to change. If they are linked together, as a group they have combined energy to work for their survival, and this energy can be harnessed to improve their situation. In fact, giving the survivors the reins to rebuild their lives is crucial and can serve as a form of therapy, as people are kept busy rather than having everything done for them by the relief agencies, which might in some cases view survivors merely as passive recipients. A top-down attitude can be disempowering for people who have already lost everything. After the initial shock of the disaster, the will of the people to survive will come through, creating a development force with huge energy, which should be harnessed as a new force for change. Of course, the less emphasis there is on the concept of collective action within the affected communities, the more fragile the concept is – and programmes such as cash-for-work schemes can break up communities by creating competition between households.

The facilitation of community-led rebuilding can strengthen communities and set them on the right path for further collective action for the benefit of all community members. Even if the affected community did not have an established community group or savings group before the disaster, the affected groups may establish ties based on their common need. The act of starting savings activities can also give the traumatised community residents a clear goal and regular activities to focus on. Community members can play an important role in all stages of a post-disaster situation, from immediate relief, surveying of affected areas, regaining livelihoods, rebuilding homes and implementing effective disaster prevention processes for the future.

The experience in the Philippines shows what can be achieved when peer communities are there to support others in the reconstruction of their lives, communities and neighbourhoods. No one community can act alone, and by collaborating with each other, through the HPFPI, and with other local community networks, and the support of professionals, local people have a chance to challenge their exclusion from decision-making processes, and 'scale-up' their influence. With such solidarity they can show that rather than the residents of informal settlements being the problem, they can be the solution. When organised and supported to plan for new and improved development options, local residents can demonstrate more effective ways of using resources and bring their own energy and commitment to survival into the recovery phase. As an organised group, they can think beyond the immediate neighbourhood to the city, and beyond recovery to the longer-term development process, addressing underlying risk factors. As a part of international networks, both Shack/Slum Dwellers International and the Asian Coalition for Housing Rights, local groups gain in credibility and then legitimacy. A shift in relationships with local authorities can occur, contributing to good urban governance, with a new style of urban management in which people are supported in meeting their demands for DRR at scale (Satterthwaite 2011: 5; UNISDR 2009). These linkages help to challenge any underlying suspicion of the urban poor and enable them to be recognised as valuable and equal partners in DRR and, by extension, development processes.

Notes

1 This section draws on Archer and Boonyabancha (2011).
2 This section draws on Carcellar *et al.* (2011).
3 The HPFPI is a duly registered national federation of close to 250 community associations of the urban and peri-urban poor in 16 cities and 16 municipalities in Luzon, Visayas and Mindanao for community-led approaches to secure tenure, upgrading, housing and basic services delivery. PACSII is the primary intermediary support institution of the HPFPI, providing 'networked intermediary support' to the emerging needs of the HPFPI on secure tenure, housing, upgrading, DRR, climate change adaptation, basic services delivery and urban liveability. Together, they form the Philippine Alliance, which organises, mobilises and networks low-income communities towards community finance and other community-driven strategies in securing tenure, implementing upgrading, housing, disaster risk management activities and urban liveability, through collaboration with government, academe and other stakeholders.
4 If the disaster-affected or disaster-threatened communities are existing active HPFPI member communities, they would have a governing structure and a core of volunteer leaders that may be tapped by the HPFPI to conduct rapid socio-economic and damage assessment surveys. For non-HPFPI-organised or non-member communities, usually the HPFPI will first settle in these communities, establish local contacts and scout for and mobilise volunteer leaders from the community to collectively plan and perform the various activities.
5 This section draws on Archer and Boonyabancha (2011).

References

ALNAP (Active Learning Network for Accountability and Performance) (2012) *Responding to urban disasters: learning from previous relief and recovery operations*, ODI.

Archer, D. and Boonyabancha, S. (2011) 'Seeing a disaster as an opportunity: harnessing the energy of disaster survivors for change', *Environment and Urbanization* 23(2): 351–364.

Carcellar, N., Rayos Co, J. and Hipolito, Z.O. (2011) 'Addressing disaster risk reduction through community-rooted interventions in the Philippines: experience of the Homeless Peoples' Federation of the Philippines', *Environment and Urbanization* 23(2): 365–381.

Davidson, C.H., Johnson, C., Lizarralde, G., Dikmen, N. and Sliwinski, A. (2007) 'Truths and myths about community participation in post-disaster housing projects', *Habitat International*, 31: 100–115.

Dodman, D. and Mitlin, D. (2011) 'Challenges for community-based adaptation; discovering the potential for transformation', *Journal of International Development*, DOI: 10.1002/jid.1772.

Hardoy, J.E., Mitlin, D. and Satterthwaite, D. (2001) *Environmental problems in an urbanizing world: Finding solutions for cities in Africa, Asia and Latin America*, Earthscan Publications.

IPCC (2012) *Managing the risks of extreme events and disasters to advance climate change adaptation (Special report on extreme events)*, Cambridge University Press.

Levine, S., Bailey, S. and Boyer, B. (2012) *Avoiding reality: Land, institutions and humanitarian action in postearthquake Haiti*, ODI.

Lyons, M. (2009) 'Building back better: the large-scale impact of small-scale approaches to reconstruction', *World Development*, 37: 385–398.

Pantuliano, S., Metcalfe, V., Haysom, S. and Davey, E. (2012) 'Urban vulnerability and displacement: a review of current issues', *Disasters*, 36: S1–S22.

Patel, S., Baptist, C. and D'Cruz, C. (2012) 'Knowledge is power: informal communities assert their right to the city through SDI and community-led enumerations', *Environment and Urbanization*, 24(1): 13–26.

Rayos Co, J.C. (2010) *Community-driven disaster intervention: Experiences of the Homeless People's Federation Philippines Inc.*, IIED.

Satterthwaite, D., (2011) *What role for low income communities in urban areas in Disaster Risk Reduction*? UNISDR and IIED.

Schilderman, T. and Lyons, M. (2011) 'Resilient dwellings or resilient people? Towards people-centred reconstruction', *Environmental Hazards*, 10(3): 218–231.

Schilderman, T. and Lyons, M. (2010) 'Building back a better Haiti', *Humanitarian Exchange Magazine*, 48, available www.odihpn.org/humanitarian-exchange-magazine/issue-48/building-back-a-better-haiti (accessed 28 December 2012).

Syukrizal, A., Hafidz, W. and Sauter, G. (2009) *Reconstructing life after the tsunami: The work of Uplink Banda Aceh in Indonesia*, IIED.

Twigg, J. (2007) *Characteristics of a disaster-resilient community: A guidance note*, DFID.

UNISDR (2009) *Global Assessment Report on Disaster Risk Reduction: Risk and Poverty in a Changing Climate*, UNISDR.

Zetter, R. and Deikun, G. (2010) 'Meeting humanitarian challenges in urban areas', *Forced Migration Review* 34.

16 The role of ecosystems in climate change adaptation

Lessons for scaling up

Helen Jeans, Judy Oglethorpe, Joanna Phillips and Hannah Reid

Everyone on Earth is dependent on nature for their well-being through the provision of water, food, shelter and clean air to breathe. Human behaviour also changes nature, thus making natural and human systems intricately connected and dependent on one another. Climate change and how people adapt to it affects these interconnected systems. This chapter explores the benefits of taking an approach to climate change adaption that integrates ecosystems into adaptation strategies to build the resilience of vulnerable people and the natural systems on which they depend.[1] It describes the principles behind ecosystem-based approaches to adaptation and explains how these can help increase people's resilience to climate change. It stresses that ecosystems function at different spatial and temporal scales and proposes that understanding these scales and working with them – in addition to social, institutional and political scales – can help ensure adaptation initiatives are effective. Likewise, it stresses that when assessing trade-offs to inform adaptation planning, ecosystems and the services they provide need to be integrated into such assessments. The chapter argues that in many instances, effective scaling up of adaptation initiatives will not be possible without considering these ecosystem-related issues in planning and implementation.

The chapter concludes by describing key lessons for successfully integrating ecosystems into adaptation strategies and programmes, including community based adapation. Particular emphasis is placed on the need to bring together skills, experience and knowledge from multiple disciplines and practices, and to share knowledge and experience from nature conservation and natural sciences to strengthen community-based adaptation strategies.

Ecosystems, people and climate change: why address these together in an integrated way?

All human communities depend on the Earth's ecosystems for their survival and well-being, but the poorest and most vulnerable people depend on them most directly. The kinds of services that ecosystems provide include *provisioning services* such as food, water, timber and fibre; *regulating services* that affect climate, floods, disease, waste and water quality; *cultural services* that provide recreational, aesthetic and spiritual benefits; and *supporting services* such as soil formation, photosynthesis and nutrient cycling (Millennium Ecosystem Assessment 2005). It is also these poor and vulnerable people who are likely to be affected most by climate change impacts because of their geographic locations in areas prone to droughts, floods and other climate change hazards, their limited financial resources, skills and technologies to cope with such hazards and their high reliance on climate-sensitive sectors such as agriculture and fishing.

Many CBA projects already use natural resources as an integral part of their approach (Reid *et al.* 2009). But this chapter proposes that effective CBA needs to more fully integrate biodiversity, ecosystem services and natural resources at the start of, and throughout the planning and implementation of CBA strategies. This will help scale up CBA and other adaptation strategies to ensure they are effective for as long as possible in the face of climate change by safeguarding ecosystem functioning and better understanding changes that will affect it. This approach to adaptation is being framed as an 'integrated approach' because it integrates some of the principles of CBA and ecosystem-based approaches to adaptation emerging from current theory and practice (Girot *et al.* 2012).

In order to address biodiversity, ecosystems and the services they provide in the context of CBA, it is necessary to recognise that just as people are being affected by climate variability and change, so too are ecosystems; they will not remain static and their functionality will be affected by climate change. Observed changes in climate have already adversely impacted certain species and ecosystems, and further changes in biodiversity are inevitable as climate change proceeds (Fischlin *et al.* 2007; Baillie *et al.* 2010). The direct impacts of climate change on the biodiversity that constitutes all ecosystems (Mitchell *et al.* 2007; Reid 2006) include:

- changes in ecosystem processes, including changes in water catchment processes and natural fire regimes;
- changes in the timing of events in the natural world, which may lead to loss of synchrony between species, for example, with insect pollinators emerging at a different time to plants flowering, making it difficult for the plant species to survive – this may include crops;
- changes in species abundance and distribution, including the arrival of new species in a place if conditions become more favourable, and the loss of species that can no longer survive there. Some species may become extinct, if they can't accommodate the changes occurring or move to more favourable conditions. This could include species that people directly or indirectly depend on;
- changes in the composition of plant or animal communities.

Climate change impacts such as the increased frequency and severity of extreme weather events like droughts, floods and storms, greater irregularity in as well as changes in physical conditions, such as snow cover, permafrost and sea level, will lead to changes in ecosystems and ecosystem functioning. Key eco-hydrological impacts of climate change on ecosystems and species are outlined by Le Quesne *et al.* (2010). Aquatic freshwater habitats and wetlands, mangroves, coral reefs, Arctic and alpine ecosystems, and cloud forests are particularly vulnerable to the impacts of climate change. The consequences of this will be experienced in different ways, such as increases in invasive species; increased human–wildlife conflict (as humans and wild species compete for the same dwindling resources such as water and grazing); increased wild fires; the emergence and increased spread of pathogens (affecting wildlife, humans and livestock); and reduced ecosystem services such as freshwater availability (FAO 2012; Fischlin *et al.* 2007). These changes can have significant social, cultural and economic consequences.

Ecosystems have limits beyond which they cannot function. When these limits are breached the ecosystem may no longer provide the services on which human communities have come to depend. There is uncertainty about the thresholds of climate change above which ecosystems are irreversibly changed and will no longer function in their current form (CBD 2009). However, we do know that ecosystems are more resilient to climate change if they are healthy and non-climate stressors such as habitat destruction, overharvesting of resources and pollution are minimised (Hansen *et al.* 2003). Given this uncertainty, adaptation strategies, including CBA, need to take a 'do no harm' precautionary approach. Promoting healthy ecosystems and reducing non-climate stressors are extremely important for maintaining ecosystem functioning and service provision for human adaptation.

Climate change will also bring indirect impacts to ecosystems and biodiversity through climate-induced changes in socio-economic drivers, including development policies, cultural values and use of land and other resources as people respond to climate change. For example, heavy investment in biofuels as a replacement for fossil fuels (to help mitigate climate change) could provide greater incentives to convert forested land to arable land. In the short term, due to their scale, scope and speed, these indirect impacts could be more damaging than the direct impacts of climate change or climate variability, especially in modified landscapes, coasts and seas (Smithers *et al.* 2008) and could reduce longer-term resilience to direct climate change impacts.

What and how can we learn from ecosystem approaches to adaptation to help increase people's resilience to climate change?

Development of ecosystem-based approaches to adaptation

There is a growing community of practice, led by the nature conservation sector, which is developing and practising ecosystem-based approaches to adaptation, often working in partnership with local communities. The Convention on Biological Diversity (CBD 2009) has defined ecosystem-based approaches to adaptation as 'the use of biodiversity and ecosystem services to help people adapt to the adverse effects of climate change'. Under this definition, ecosystem-based approaches to adaptation use a range of opportunities for the sustainable management, conservation and restoration of ecosystems to provide services that help people adapt to the impacts of climate change. Examples of ecosystem-based approaches to adaptation include (Munroe *et al.* 2011; CBD 2009):

- defence of coastlines in the face of more extreme storms by maintaining and/or restoring coastal vegetation such as mangroves. The vegetation reduces the strength of waves before they reach the shore and therefore reduces coastal flooding and erosion;
- sustainable management of wetlands and floodplains for maintenance of water flow and quality in the face of changes in precipitation and streamflow, acting as floodwater reservoirs to reduce downstream flooding, and providing important stores of water in times of drought;
- conservation and restoration of forests and natural vegetation to stabilise slopes and regulate water flows, reducing flash flooding and landslides as rainfall levels and intensity increase;
- establishment of healthy and diverse agricultural and agroforestry systems to secure food provision in changing climatic conditions by maintaining genetic diversity of crops and livestock and using indigenous knowledge of crop and livestock varieties.

Ecosystems function at multiple levels and across a range of different sectors such as water, agriculture, energy, transport, health and conservation. Ecosystem-based approaches to adaptation must therefore work with a wide range of stakeholders at all levels and from different disciplines to promote integrated solutions to climate change including through multi-agency cooperation. This includes working with local and indigenous communities. For example, see Box 16.1, which summarises a programme of work being implemented by the WWF in Belize, which takes an ecosystem-based approach to adaptation and works with a wide range of stakeholder groups.

Principles for ecosystem-based approaches to adaptation

Core principles for ecosystem-based approaches to adaptation have been developed by Andrade *et al.* (2011) for use when undertaking national adaptation planning, by financial institutions, and in project and research design. These principles have been developed from lessons from the field and from well-established principles for taking an ecosystem approach, which were adopted by the Convention on Biological Diversity in 2000 and endorsed by the World Summit on Sustainable Development in 2002. Andrade *et al.* (2011) state that an ecosystem-based approach to adaptation:

- is about promoting the resilience of both ecosystems and societies;
- promotes multi-sectoral approaches;
- operates at multiple geographical and temporal scales;

Box 16.1 Mangrove restoration in Belize

National-level mapping of coastal mangrove coverage for Belize was carried out to improve understanding of the marine and coastal ecosystems, identify potentially resilient mangrove habitats and also identify those coastal areas most vulnerable to sea-level rise and storm surges. This WWF programme then undertook work to protect and restore mangroves and reduce non-climate impacts on the marine ecosystem. Focusing initially on development hot-spots and aquaculture farms, the WWF has been working with teams of local people and school children to plant over 17,000 red mangrove seedlings. Planting at shrimp farms helps remove pollution and farm effluents from the water, thus improving the lagoon. Mangrove restoration initiatives tested at the local level with private developers have been scaled-up at the national level with governmental partners, with a view to initiating national government ownership of the process. The WWF is also advocating for the designation of private mangrove reserves in Belize to protect remaining virgin mangrove stands. In Placencia, for example, many large developments have set aside between 10 and 60 per cent of their properties as mangrove reserves, but these blocks remain unmanaged and unmonitored. The WWF has been in dialogue with property owners, developers and the Belize Association of Private Protected Areas to get formal designation and management of these private reserves in order to safeguard their ecosystem services and thus develop a scaled up strategy for developing the coastal zone.

- integrates flexible management structures that enable adaptive management;
- minimises trade-offs and maximises benefits with development and conservation goals to avoid unintended negative social and environmental impacts;
- is based on best available science and local knowledge, and fosters knowledge generation and diffusion;
- is about resilient ecosystems, and using nature-based solutions in the service of people, especially the most vulnerable;
- is participatory, transparent, accountable and culturally appropriate, and actively embraces equity and gender issues.

These principles emphasise that ecosystem-based approaches are most appropriately integrated into broader adaptation and development strategies. When done effectively this can be a very cost-effective way to secure adaptation and development benefits that also generate social, economic and cultural co-benefits, including disaster risk reduction, livelihood sustenance and food security, carbon sequestration and sustainable water management (CBD 2009; Colls *et al.* 2009).

Many of these principles are also important and increasingly addressed in the context of CBA, as some of the chapters in this book describe. For example, to be effective CBA needs to be addressed not just at the local level, but also embedded in wider government and institutional processes that can help mainstream CBA into wider adaptation planning processes. Long-term as well as short-term temporal considerations are also important for CBA in order to avoid eventual maladaptation, and good CBA inevitably involves the consideration of trade-offs in order to make the best decisions regarding how to invest limited resources and what to prioritise in terms of adaptation actions.

We argue here that when considering issues related to scale and trade-offs in the design and implementation of CBA initiatives, it is essential to include an ecosystem component in this planning. This is particularly important in the context of scaling up CBA, because attempts to scale up CBA will likely struggle or even fail if the only geographical and temporal scales considered are those of a social, developmental nature and exclude ecological considerations. Consideration needs to be given to both how ecosystems can help people to adapt, and also what needs to be done to help natural systems to adapt at different scales. CBA successes may also be compromised if trade-offs fail to incorporate the full value that ecosystem services and biodiversity can provide in the context of adaptation.

Spatial scales

Ecosystems function at different and often nested scales, from local (for example, the catchment of a small pond) to very large (e.g. an international river basin). Very often their boundaries do not correspond with political boundaries (e.g. village, district, provincial or national boundaries). It is important that adaptation planning takes into account and ensures harmonisation between scales of critical ecosystem function and political and social scales of intervention. For example, if neighbouring villages facing a drought develop individual adaptation plans that call for sinking wells, without collectively considering the capacity of the aquifer to provide water for all villages, benefits may be short-lived if the aquifer is over-pumped in the longer term. The result could be an exacerbation of both human and ecosystem vulnerability – resulting in maladaptation.

Operating at multiple scales simultaneously is important and can be done by adopting a systems approach (a holistic approach that takes into account interactions and interdependencies at different levels) rather than an individual project at a single scale. For example, local planning for sinking wells should be integrated into broader river basin planning.

Pastoralists in drought-prone landscapes also provide a good example of such a systems approach in action. Many have well-developed strategies to manage risk and enhance resilience to cope with droughts that may be applied as climate variability increases. These also remind us of the importance of understanding and learning from traditional knowledge and practices.

Working at an ecosystem scale may require multi-stakeholder and regional or trans-boundary collaboration to secure fair and effective policies and management of the resource or ecosystem (e.g. river basin) that a community depends upon. This means working beyond the community level and engaging in governance systems and political processes that don't necessarily share boundaries with ecological systems and processes. This can provide opportunities to develop adaptive capacity beyond the local level by promoting collaborative governance and generating knowledge about how best to apply such systems approaches. Emerging policies can provide enabling environments for adaptation and opportunities to learn from traditional management systems in vulnerable ecosystems where coping strategies are inherently resilient. This can encourage scaling up based on sound ecosystem and equitable governance practices. Box 16.2, for example, shows the need to address different geographic, social and ecological scales to achieve success at the local level, in work undertaken by Mercy Corps to build the resilience of slum dwellers in low-lying parts of northern Jakarta, Indonesia.

Box 16.2 Climate change, adaptation and development in Jakarta

Many of the urban poor of Jakarta are squatters living in the low-lying northern part of the city along river banks and canals. Inadequate piped water provision results in much of the city's population using shallow groundwater for their water needs, causing subsidence. Sewerage is largely inadequate and many wells are contaminated by faecal matter. There is no citywide solid waste collection scheme, and much garbage is dumped into streets and canals and scavenged by garbage pickers.

In 2006 Mercy Corps began implementing the Healthy Places Prosperous People programme, funded by the International Development Research Centre. The goal was to improve lives and livelihoods of people living in a slum in low-lying northern Jakarta. One pillar of this programme was securing water supplies and improving sanitation and solid waste management. The project had just started when, in February 2007, Jakarta experienced its worst recorded flood, which inundated large parts of the city, killed at least 57 people and sent over 422,000 fleeing their homes. Those hardest hit were the poor people living in wooden houses along the riverbanks. Immediately after the flood, many thousands of people became ill with flood-related diseases, especially diarrhoea and dengue fever. The flood damaged many of the investments that Mercy Corps had made in the communities, and their emergency services were overwhelmed.

The flood was a result of heavy rain like that expected with increasing climate variability, exacerbated by deforestation on higher land south of the city, which caused rapid water runoff, and waterways clogged with debris which impeded drainage. Mercy Corps realised that it would not be possible to build the resilience of the communities they worked with and help them to adapt to future climate change by only working on local projects. A larger-scale, holistic, integrated and multi-disciplinary approach was needed. They therefore began assessing problems in the entire region, not just the poor communities and slum areas where they had a physical presence.

Since the 1960s, mountain areas outside of the city have been increasingly deforested to make way for villas and other building developments. More recently, towns have expanded onto cleared agricultural land outside the city. Green spaces, which used to absorb rain water, have all but disappeared as Jakarta has joined with neighbouring municipal areas. Canals in the downstream Jakarta area provided the only exit for excess water, but increased amounts of solid waste from the growing population and insufficient waste collection meant garbage was accumulating in the canals, so rushing waters quickly spilled into flood-prone areas of the city.

Living in these low-lying, subsiding, flood-prone areas were vulnerable people, often in slums. These vulnerable people, therefore, bore the brunt of 50 years of deforestation, disorganised urban sprawl, creaking infrastructure and inadequate social services. They were the victims, not the causes of the floods. And their vulnerability will likely increase as land subsidence continues, more extreme weather events occur as climate variability increases, and sea-level rise advances in the future due to climate change.

Mercy Corps, a humanitarian and development organisation, realised that effective adaptation meant collaborating with conservation groups working in the mountain areas to conserve the forests, and also with local municipal authorities on garbage management, runoff and canals. Mercy Corps has since gone on to develop a guide for its staff on climate change adaptation and resilience-building programmes for cities, which outlines the importance of taking a multi-disciplinary, holistic approach and embedding programmes in government structures in order to make them sustainable (Mercy Corps 2011).

Source: World Bank (undated); Jim Jarvie, Mercy Corps (pers. comm.) and Shaun Martin, WWF US (pers. comm.).

Temporal scales

Temporal scales are also important in adaptation. Mangroves provide a good example of the long time frames that need to be considered in adaptation and land use planning. As previously outlined, mangroves can buffer the effects of storm surges, providing protection to vulnerable coastal communities and infrastructure, as well as important fisheries. In the longer term, though, mangroves will be affected by sea-level rise and risk being submerged and lost unless they can move inland, keeping pace with rising water, if the topography is suitable. In some countries, such as Vietnam and Australia, communities are setting aside areas behind existing mangroves so they have space to move inland and thus continue this important ecosystem function. It can take time for climate impacts on ecosystems to become visible, although natural infrastructure is sometimes surprisingly fast in establishing itself and starting to provide services and multiple benefits.

Minimising trade-offs and maximising benefits with development and conservation goals

The use of land and natural resources will need to be adjusted to changing climate conditions. This will inevitably involve trade-offs in many parts of the world where demand and competition for natural resources and ecosystem services increase as climate change advances. In some places this will be exacerbated by the concurrent advance of development and population growth. Such trade-offs will vary depending on the given context. While ecosystem-based approaches to adaptation cannot avoid this, their application can contribute towards fully informed and improved decision-making processes that bring ecosystem issues into the equation and ensure a full assessment of costs and benefits occurs rather than simple, short-term economic analyses that fail to capture the full economic and non-economic benefits that ecosystem services and effective biodiversity management can provide.

This is particularly important in the management of fresh water, a resource that is coming under increasing pressure. Since there is still a high degree of uncertainty about future climate change in some areas, particularly for precipitation trends, scenario planning that takes into account both development and climate scenarios can help to assess potential future ecosystem impacts and make more comprehensively reasoned and phased decisions (Le Quesne *et al.* 2010).

Opportunities and principles for integrating ecosystems into CBA strategies

Combining insights from conservation and CBA communities of practice and the science that supports them enables a wider range of methodologies, knowledge and skill sets to be applied in addressing the challenge of adaptation. It creates opportunities for incremental improvements in conservation, development and climate change practices, as well as a potentially transformative reframing of approaches to the way we work.

CBA practitioners have prioritised the following in their work:

- developing methodologies that recognise human rights, gender and equality;
- building on the priorities and capacities of local people;
- identifying different traditional knowledge;
- strengthening the social aspects of monitoring and evaluation; and
- ensuring that interventions build rather than undermine social capital (a key constituent of adaptive capacity).

Conservation practitioners have prioritised the following in their work:

- developing approaches for understanding the impact of climate and non-climate threats on biodiversity, ecological systems and ecosystem services such as pollination, climate regulation, genetic diversity and ecological connectivity, including potential tipping points;
- working at multiple scales, taking ecosystem processes into account;
- incorporating ecosystem goods and services into community-led adaptation strategies and building ecosystem resilience at multiple scales;
- incorporating the latest science on climate change impacts on social and environmental systems in participatory vulnerability analyses;
- building environmental integrity into monitoring and evaluation systems;
- ensuring that interventions build rather than undermine natural capital (another key constituent of adaptive capacity); and
- helping facilitate adaptation of natural systems working at multiple scales.

Learning together

By learning from each other these communities of practice will be more able to develop effective adaptation strategies and provide cost-effective alternatives to top-down, hard infrastructure-based approaches to adaptation, such as concrete sea walls or canals. While in some places hard infrastructure may be the only solution left, in many places locally appropriate environmentally robust solutions will strengthen and build the resilience and long-term adaptive capacity of communities and the ecosystems they depend on, while providing multiple co-benefits and retaining options for the future in the face of uncertainty.

Collaborative partnerships that facilitate social learning are becoming more important than ever before in order to deal with the scale, complexity, urgency and uncertainty of a climate change constrained world. Local communities and development, disaster risk reduction and conservation communities of practice have complementary knowledge, experience, skills and tools from top-down and bottom-up approaches which can contribute to the development of an integrated adaptation field of practice. New and expanded partnerships are needed to promote transdisciplinary, multi-sectoral and multiple-scale approaches to adaptation. Joint activities, dialogue and social learning are necessary to build bridges between communities of practice and with local communities and decision-makers to develop a sound understanding of the opportunities, challenges and limits of different adaptation approaches. This will be a challenge in the context of integrating ecosystems into CBA better, because in practice, local climate change project activities are funded and implemented by both conservation and development organisations, which at times have very different skill sets, knowledge bases, values and ideologies. But it is a challenge that needs to be addressed and overcome.

Community-based natural resource management (CBNRM) initiatives in Africa (Murphree 1998), for example, have shown that responsibilities for natural resources management are more effectively implemented when management is closer to the social grouping affected or benefiting. As with CBNRM, both CBA and ecosystem-based approaches to adaptation should support decentralisation to the lowest appropriate level, avoiding top-down approaches that may not take local interests and ecosystems fully into account. Such approaches could support resource-poor communities that may be forced to adopt short-term solutions that prove maladaptive in the medium and long term; for example, forest communities who have no access to alternative fuel and therefore cut down trees for their fuel needs. In this case supporting local communities to access fuel in ways that manages the forest resource for ecological integrity could support both community resilience and the resilience of the forest ecosystem in the medium to long term.

An important principle of CBA is to take account of the different vulnerability of women, children and vulnerable groups to climate variability and change. In poor rural communities, these groups are often dependent on ecosystem services, and often play a significant role in natural resource management. Supporting these groups to make adaptive choices that support the ecosystems on which they depend would contribute to the medium- and long-term resilience of these groups and ecosystems which are also undergoing change.

Like CBA, ecosystem-based approaches to adaptation benefit from traditional knowledge and local practices that have been developed to adapt to climate variability (CBD 2009). For example, more than half of households surveyed in a case study from Guangxi Province in China still use local landraces of maize and rice because they taste good, are better adapted to mountainous and barren land, and have drought and lodging resistance (to protect from wind force). Most of these farmer-improved landraces survived the big spring drought in 2010, while most of the modern hybrids were lost. And in the Bolivian Andes, farmers are using local potato varieties in response to new pests and water shortages (Swiderska *et al.* 2011). It is recognised that some of the changes that are being experienced now and are likely in the future go beyond living memory and may require a combination of endogenous innovation and exogenous inputs. Local observation of changes in weather, hazards and impacts, along with scientific projections of future changes and scenario planning, can contribute to assessing current and future vulnerability of people and ecosystems and planning responses.

Some key lessons for successfully integrating ecosystems into adaptation strategies and programmes, including CBA, are as follows:

- As well as working the local community level, think and plan at the spatial scales that ecosystems operate at, and engage with institutions operating at these other scales (e.g. the watershed management scale) to create an enabling institutional environment that can address ecosystem functions.
- Integrate endogenous (internal) and exogenous (external) knowledge so that local and traditional knowledge on natural resource/ecosystem management for livelihoods/human well-being is supported by appropriate exogenous technology.
- Maintain ecosystem services at multiple scales (e.g. replanting deforested slopes, managing water resources on a catchment-wide scale with multi-stakeholder participation; and restoring and replanting mangroves).
- Increase ecosystem resilience (e.g. by removing non-climate stresses, and promoting ecosystem management and restoration).
- Facilitate ecosystem adaptation (recognising that ecosystems and natural resources will not remain stationary in the face of climate change).
- Promote no/low regret, locally appropriate nature-based solutions.
- Promote partnerships that bring together local communities and decision-makers with development, conservation and disaster risk reduction communities of practice.
- Ensure the full extent of ecosystem services and values is included in any assessment of costs and benefits for adaptation planning.
- Monitor and manage ecosystem change to support ongoing human and ecosystem resilience and adaptation.

Scaling up CBA: the importance of incorporating ecosystem-based approaches

CBA is usually approached primarily from a social or developmental perspective, which may or may not look at how ecosystems and the services they provide can help communities adapt. This chapter argues, however, that effective CBA needs to address issues relating to biodiversity, ecosystem services and natural resources right at the start of any assessment, planning and action on CBA, and that successful upscaling of CBA could fail if this does not occur. CBA practitioners, for example, sometimes see ecosystems and ecosystem services as static, but this could render efforts at scaling up useless. A scaled up mangrove planting operation that fails to consider the impact that sea-level rise will have on the mangroves will ultimately benefit no-one in the longer term. Likewise, scaling up CBA projects that do not consider ecological knowledge relating to the minimum size of ecological functioning units required for effective forest functioning could fail and potential resulting adaptation benefits would be lost if the forest system is fragmented to the extent that it loses its ecological integrity.

Good CBA is primarily bottom-up and grounded in community values, coping strategies, priorities and decision-making structures. But to scale up CBA these need to be considered within the context of larger structures, such as National Adaptation Programmes of Action or city development plans. This chapter argues that just as CBA needs to be integrated into top-down government-led policy and planning processes to be scaled up effectively, it also needs to be integrated into larger-scale issues relating to ecosystem functioning such as watersheds, natural coastal defence systems, sustainable forest management plans, and so on. Without this, scaling up could end up being maladaptive in many natural environments, and even many urban environments, for example if city water management plans do not consider wider watershed management issues.

Note

1 The Intergovernmental Panel on Climate Change includes both social and ecological components in its definition of resilience, which we adhere to in this chapter: 'The ability of a social or ecological system to absorb disturbances while retaining the same basic structure and ways of functioning, the capacity for self-organisation, and the capacity to adapt to stress and change.'

References

Andrade, A., Córdoba, R., Dave, R., Girot, P., Herrera, F.B., Munroe, R., Oglethorpe, J., Paaby, P., Pramova, E., Watson, E. and Vergar, W. (2011). *Draft Principles and Guidelines for Integrating Ecosystem-based Approaches to Adaptation in Project and Policy Design: A discussion document.* IUCN-CEM, CATIE.

Baillie, J.E.M., Griffiths, J., Turvey, S.T., Loh, J. and Collen, B. (2010). *Evolution Lost: Status and trends of the world's vertebrates*. Zoological Society of London.

CBD (2009). *Connecting Biodiversity and Climate Change Mitigation and Adaptation: Report of the Second* Ad hoc *Technical Expert Group on Biodiversity and Climate Change.* Secretariat of the CBD.

Colls A., Ash, N. and Ikkala, N. (2009). *Ecosystem-based Adaptation: A natural response to climate change.* IUCN.

FAO (2012). *Wildlife in a Changing Climate.* Food and Agriculture Organization.

Fischlin, A., Midgley, G.F., Price, J.T., Leemans, R., Gopal, B., Turley, C., Rounsevell, M.D.A., Dube, O.P., Tarazona, J. and Velichko, A.A. (2007). Ecosystems, their properties, goods, and services. *Climate Change 2007: Impacts, Adaptation and Vulnerability. Contribution of Working Group II to the Fourth Assessment Report of the Intergovernmental Panel on Climate Change,* M.L. Parry, O.F. Canziani, J.P. Palutikof, P.J. van der Linden and C.E. Hanson, Eds., Cambridge University Press.

Girot, P., Ehrhart, C. and Oglethorpe, J. (2012). Integrating community and ecosystem-based approaches in climate change adaptation responses.. http://elanadapt.net/sites/default/files/siteimages/elan_integratedapproach_15_04_12_0.pdf (accessed 15 July 2013).

Hansen, L., Biringer, J.L. and Hoffman, J. (2003). *Buying Time: A user's manual to building resistance and resilience to climate change in natural systems.* World Wildlife Fund.

Le Quesne, T., Matthews, J.H., Von der Heyden, C., Wickel, A.J., Wilby, R., Hartmann, J., Pegram, G., Kistin, E., Blate, G., de Freitas, G.K., Levine, E., Guthrie, C., McSweeney, C. and Sindorf, N. (2010). *Freshwater Ecosystem Adaptation to Climate Change in Water Resources Management and Biodiversity Conservation.* World Bank.

Mercy Corps (2011). *Urban Climate Change Adaptation for Mercy Corps Program Designers and Implementers: Lessons from ACCCRN in Indonesia.* Mercy Corps.

Millennium Ecosystem Assessment (2005). *Ecosystems and Human Well-being: Synthesis.* Island Press.

Mitchell, R.J., Morecroft, M.D., Acreman, M., Crick, H.Q.P., Frost, M., Harley, M., Maclean, I.D.M., Mountford, O., Piper, J., Pontier, H., Rehfisch, M.M., Ross, L.C., Smithers, R.J., Stott, A., Walmsley, C.A., Watts, O. and Wilson, E. (2007). *England Biodiversity Strategy: Towards Adapation to Climate Change.* Final report to Defra.

Munroe, R., Doswald, N., Roe, D., Reid, H., Giuliani, A., Castelli, I. and Moller, I. (2011). *Does EbA Work? A review of the evidence on the effectiveness of ecosystem-based approaches to adaptation.* Cambridge Conservation Initiative.

Murphree, M. (1998) *Synergizing Conservation Incentives: From local global conflict to compatibility.* Institute for Development Policy and Management.

Reid, H. (2006). Climatic change and biodiversity in Europe. *Conservation and Society* 4(1): 84–101.

Reid, H., Alam, M., Berger, R., Cannon, T., Huq, S. and Milligan, A. (2009). Community-based adaptation to climate change: an overview. *Community-based adaptation to climate change: Participatory Learning and Action* 60: 11–33.

Smithers, R.J., Cowan, C., Harley, M., Hopkins, J.J., Pontier, H. and Watts, O. (2008). *England Biodiversity Strategy: Climate change adaptation principles – conserving biodiversity in a changing climate.* DEFRA.

Swiderska, K., Song, Y., Li, J., Reid, H. and Mutta, D. (2011). *Adapting Agriculture with Traditional Knowledge.* IIED.

World Bank (undated). *Jakarta case study overview. Climate change, disaster risk and the urban poor: Cities building resilience of a changing world.* http://siteresources.worldbank.org/INTURBANDEVELOPMENT/Resources/336387-1306291319853/CS_Jakarta.pdf (accessed 4 April 2012).

Index

Page numbers in *italics* denote tables, those in **bold** denote figures.

CPSIA information can be obtained
at www.ICGtesting.com
Printed in the USA
BVHW040337090419
544967BV00010B/260/P

9 780415 623698